From a to α

Yeast as a Model
for Cellular Differentiation

ALSO FROM COLD SPRING HARBOR LABORATORY PRESS

Related Manuals

Methods in Yeast Genetics: A Laboratory Course Manual, 2005 Edition

Other Available Titles

The Coiled Spring: How Life Begins
Genes & Signals
A Genetic Switch, Third Edition, Phage Lambda Revisited
Landmark Papers in Yeast Biology
Lateral DNA Transfer: Mechanisms and Consequences

From a to α

Yeast as a Model
for Cellular Differentiation

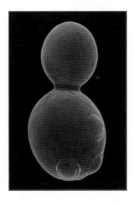

Hiten D. Madhani

University of California, San Francisco

COLD SPRING HARBOR LABORATORY PRESS
Cold Spring Harbor, New York

From a to α
Yeast as a Model for Cellular Differentiation

Publisher	John Inglis
Editorial Director and	
Acquisition Editor	Alexander Gann
Development Director	Jan Argentine
Production Manager	Denise Weiss
Developmental Editor	Siân Curtis
Project Coordinator	Inez Sialiano
Permissions Coordinator	Maria Fairchild
Production Editor	Kathleen Bubbeo
Compositor	Compset Inc.
Desktop Editors	Lauren Heller and Susan Schaefer
Cover Designer	Paula Goldstein

Front Cover Artwork: Scanning electron micrograph of *Saccharomyces cerevisiae*. Credit: J. Forsdyke/ Gene Cox/Photo Researchers, Inc.

Library of Congress Cataloging-in-Publication Data

Madhani, Hiten D.
  From a to [alpha] : yeast as a model for cellular differentiation /
Hiten D. Madhani.
    p. cm.
  Includes index.
  ISBN 978-087969737-2 (hardcover : alk. paper) -- ISBN 978-087969738-9
(pdk. : alk. paper)
  1. Cell differentiation. I. Title.
  [DNLM: 1. Saccharomyces cerevisiae--cytology. 2. Cell Differentiation. 3. Cell
Cycle. 4. Models, Biological.   QU 375 M181f 2006]

   QH607.M33 2006
   571.8'636--dc22

                        2006001765

All Cold Spring Harbor Laboratory Press publications may be ordered directly from Cold Spring Harbor Laboratory Press, 500 Sunnyside Blvd., Woodbury, New York 11797-2924. Phone: 1-800-843-4388 in Continental U.S. and Canada. All other locations: (516) 422-4100. FAX: (516) 422-4097. E-mail: cshpress@cshl.edu. For a complete catalog of all Cold Spring Harbor Laboratory Press publications, visit our World Wide Web Site http://www.cshlpress.com/

Dedicated to the memory of Ira Herskowitz

CONTENTS

PREFACE

HOW ARE THE PRINCIPAL COMPONENTS OF CELLS—their genes, proteins, and membranes—modulated to produce distinct types of cells? This is perhaps the most fundamental question of modern biology and impinges on our understanding of development of multicellular organisms, evolution, and human disease.

The chapters in this book form a story about what scientists have learned about the process of cellular differentiation from studies of baker's yeast, *Saccharomyces cerevisiae*. Much more closely related to humans than bacteria, this organism has served as a simple and experimentally tractable "model" for human cells—a notion that has been borne out by research carried out during the last 25 years. Processes from gene transcription to protein secretion to cell signaling happen essentially the same way in yeast as in human cells. This suggests that the fundamental aspects of eukaryotic cells evolved more than a billion years ago and that they were then inherited by all of the descendants of that original cell, including us. Of course, complexities were added to this foundation over time, but the core of cell biology has been conserved during evolution.

The goal of this book is to introduce the knowledge gained from studies of yeast to those interested in how cells acquire distinct characteristics from each other, a process called cellular differentiation. Although this information can be found in encyclopedic textbooks, the aim here is to cover these same concepts, but concisely in the context of a single narrative.

To make this story succinct enough for the beginner to digest, I have made no attempt to describe the experiments that led to the knowledge described. Those interested will find references to relevant review articles at the end of each chapter. To help readers connect the concepts from yeast to other organisms, within the chapters I have included "boxes" that describe the relevance of the material in the chapter to problems in higher organisms and human disease.

Many kind people helped with this effort. The help, encouragement, and perspective of Alex Gann, who approached me about this project in the first place, was essential. The following colleagues provided much-needed advice: Angelika Amon, Jim Broach, Hana El-Samad, Shiv Grewal, Jim Haber, Sandy Johnson, David Julius, Hay-Oak Park, Terry Shock, George Sprague, David Stillman, Annie Tsong, and Brian Tuch. Advice on writing from Mark Ptashne was particularly helpful. Leemor

Joshua-Tor generated eye-pleasing figures of high-resolution structures. Also indispensable were Siân Curtis, Susan and Hans Neuhart, Inez Sialiano, Kathy Bubbeo, Denise Weiss, Jan Argentine, and others at CSHL Press. Finally, none of this would have been possible without the love and support (and advice) of Suzanne Noble.

Hiten Madhani
August 2006
San Francisco

FOREWORD

THESE DAYS A MOLECULAR BIOLOGIST IS CONFRONTED by two seemingly discordant narratives.

On the one hand, there is a unity, even a sameness, that extends to many biological processes across disparate organisms. We bear essentially the same enzymes as *Drosophila*, for example, and our differences are largely accounted for by the different ways in which those common enzymes are deployed during development.

On the other hand, we are evermore in danger of drowning under the weight of facts. The typical molecular biology seminar can quickly bewilder even an interested listener because it is so hard to keep track of what all the three-lettered names (Sxi, Ime, Opb, etc.) stand for and what the corresponding gizmos do. The problem is compounded by an artifact of our success: In the "classical" era, we typically began with a distinct mutant phenotype and tried to figure out its molecular basis. But now our technology is so powerful that we can start with the molecules, build a picture involving complex interactions and reactions, and yet not be sure of the biological importance of what we have found.

Is there a promised land? Will we some day gaze at sequences, apply a super systems analysis using our new computers, and sit back and watch fall out a description of how things work? Maybe so. But what do we do in the meantime?

Here the book in your hands should provide some solace. You will find a series of short chapters, each of which deals with the description of molecular interactions that solve a specific biological problem. All of these solutions are as found in a single organism, the yeast known as *Saccharomyces cerevisiae*. And so a wide array of genetical, cell biological, biochemical, and physical tests can be used to "surround" each problem and provide an answer in a clear biological context.

"The most important product of basic research," someone once said (I believe it was Dudley Herschbach), "is ideas." That message is reinforced here: As Madhani outlines in his introductory chapter, experimental manipulation of this single-celled organism has played an important role in developing many of our ideas concerning gene regulation; signaling within and between cells; the generation of differences between cells (differentiation); and so on. If you are like me, you will have a better chance of grasping these ideas as presented in this little book than if you take on the typical massive text. The more we learn about complex organisms in light of what we have learned (or thought we have learned) from the study of simpler organisms, the more Karl Popper's insight that "classical models tell us more than we can at first know" proves its aptness. There is every reason to believe that the study of yeast, and the messages of this book, will continue to be gifts that keep on giving.

Mark Ptashne

CHAPTER ONE

WHY READ A BOOK ABOUT YEAST?

Cells are the fundamental units of life. Multicellular organisms such as ourselves are made of many different kinds of specialized cells. For example, neurons, found throughout the body, conduct nerve impulses, sometimes over substantial distances; cardiomyocytes, heart muscle cells, contract to produce heartbeats; and enterocytes, which line the gut, transport nutrients into the bloodstream. The specific properties of individual cells determine how a tissue, an organ, and, ultimately, an entire organism function.

The idea that multicellular organisms are built of specialized cells was originally inferred from microscopy. Within a given organ or tissue, a discrete number of recurring cell morphologies were seen, which led to the idea that there were a limited number of "cell types" from which organs and tissues were built. It is now estimated that the trillion or so cells in the human body comprise some 200–300 cell types, but what gives different cells their distinctive properties and how do all of the different cell types arise from a single fertilized egg?

Molecular Biology of Cell Type

In recent years, great strides in our understanding of how different cell types arise have been made. The major discoveries can be organized into several principles.

1. The presence of particular "master regulatory proteins" specifies a given cell type. These proteins function by controlling the transcription of a set of genes. Regulatory proteins can be mixed and matched such that different combinations of regulatory proteins produce different cell types.

2. Proteins encoded by these gene sets impart onto a given cell type all of its unique properties. Thus, for example, it is the expression of particular

sets of genes that endows a neuron with the ability to conduct nerve impulses, a cardiomyocyte the ability to contract during a heartbeat, an enterocyte the ability to transport nutrients, and so on.

3. In addition to setting up the intrinsic properties of cells (such as shape and size), the master regulatory proteins also determine how a given cell type communicates with other cells in its environment. Thus, different cell types send out distinct signals and these, in turn, can be interpreted differently by different cell types.

4. As cells divide during development, new cell types are generated. One way that this occurs is through so-called "asymmetric" cell divisions in which a cell divides to produce two cells that are of different cell types from each other. This mechanism is fundamental to the generation of cell diversity because all of the cells of a multicellular organism are descendents of a single cell—namely, the fertilized egg. There are also other important mechanisms that generate cell types; these require communication between cells.

Yeast as Model

Much of the progress in understanding how different cell types arise came not from studying human beings directly, but from tackling far simpler organisms. One of these so-called "model organisms" is baker's yeast, also known as *Saccharomyces cerevisiae* (I will use "yeast" and "*Saccharomyces cerevisiae*" interchangeably for the remainder of the book). This single-cell organism exists as one of a handful of cell types, and each cell type is easy to manipulate experimentally. Consequently, our understanding of the principles outlined above is considerably deeper for yeast than for any other organism. It is this fact—that yeast is the single best-understood example of cell specialization—that is the motivation behind this book. The insights learned from yeast also apply to humans.

In this book, we will focus on three of the cell types of yeast, which are introduced in Figure 1-1. To orient the reader, Figure 1-2 shows the analogous view of the more familiar human reproductive life cycle. Two of the cell types in yeast, called **a** and α, are the equivalents of egg and sperm of animals. They can mate with each other to produce a third cell type: the **a**/α cell type. Under certain environmental conditions, **a**/α cells undergo meiosis to produce haploid **a** and α gametes. These gametes are packaged as spores that can lie dormant for months until conditions become favorable for growth. Once nutrients become available, they "germinate" into **a** or α cells that are capable of dividing (Fig. 1-1). Spores can be considered another cell type that is a quiescent intermediate state between the **a**/α cell type and the **a** or α cell types. Our focus in this book will be on the three cell types called **a**, α, and **a**/α.

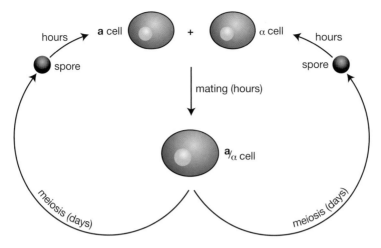

Figure 1-1. Yeast cell types in the context of the sexual cycle.

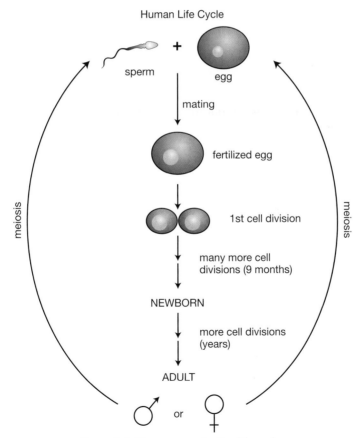

Figure 1-2. Human reproductive life cycle.

Preview of the Book

To organize our understanding of these three yeast cell types, I have divided the book as follows. Chapter 2 describes the properties that distinguish one cell type from another. Chapter 3 is focused on the master regulatory molecules that specify the **a**, α, and **a**/α cell types by controlling the expression of different sets of genes. Chapter 4 describes molecules called pheromones that **a** and α cells use to signal each other to stop dividing and initiate the process of mating. Chapter 5 explains how mating pheromones are detected by specific receptors and how this leads to mating between **a** and α cells. In Chapter 6, an example of an asymmetric cell division is described, and Chapter 7 details a control mechanism necessary for this process called "gene silencing." Chapter 8 introduces a difference in the site of cell division chosen by distinct cell types and explains how this difference arises. Finally, Chapter 9 describes our understanding of how cell types are specified in two other fungal species.

Throughout the text I have attempted to introduce the necessary background for the reader before describing a particular concept. However, a basic understanding of the nature of a eukaryotic gene and the organization of a typical eukaryotic cell will be assumed. This information is available in introductory cell biology textbooks such as *Molecular Biology of the Cell*, by Alberts et al.

CHAPTER TWO

YEAST CELL TYPES

This chapter will acquaint the reader with the ways in which the **a**, α, and **a**/α cell types differ from one another. We will learn about the chemical messengers or pheromones that **a** and α cells send out to attract each other for mating and the receptors that **a** and α cells use to sense these pheromones. We will also discuss in detail the mating process itself. Next, we will cover two processes that occur only in the **a**/α cell type: meiosis and sporulation. When not reproducing sexually, all three of these cell types can proliferate in an asexual manner, and we will learn that the cell types differ in some aspects of proliferative growth. The chapter will end with a summary of the properties that distinguish the **a**, α, and **a**/α cell types. The remainder of the book will focus on the molecules and mechanisms that give rise to these cell type–specific properties.

The **a** and α Cell Types Are Haploid, whereas the **a**/α Cell Type Is Diploid

Yeast is a eukaryote. Like other eukaryotic cells, yeast cells each have a nucleus and cytoplasm, as well as cytoplasmic organelles such as mitochondria, the endoplasmic reticulum (ER), and the Golgi apparatus. These organelles function as they do in other eukaryotes such as animals and plants. The plasma membrane of yeast is enclosed in a protective cell wall made of carbohydrate and protein.

Within the nucleus of the cell, the yeast genome contains about 6000 genes that are distributed among 16 chromosomes. Each chromosome contains a single, linear piece of DNA. The **a** and α cell types are haploid, which means that they contain one copy each of the 16 chromosomes. The **a**/α cell type is diploid, which means that it contains two copies of each chromosome.

a and α Cells Secrete Cell Type–specific Mating Pheromones That Are Sensed by Cell Type–specific Receptors

a and α cells mate only with each other—for example, an **a** cell will mate with an α cell but not with another **a** cell. For this reason, the **a** and α cell types are also referred to as "mating types." Each mating type produces a different mating pheromone (Fig. 2-1). Mating-type **a** cells produce a pheromone called **a**-factor, a small peptide, and mating-type α cells produce a pheromone called α-factor, a small peptide whose structure is distinct from that of **a**-factor.

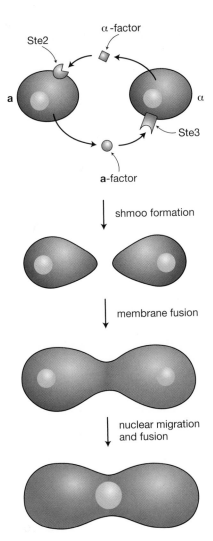

Figure 2-1. Yeast of opposite mating types signal each other with peptide pheromones in order to mate with each other.

The **a** and α cell types recognize the pheromone produced by their opposite mating type via specific receptors on the cell surface. Thus, **a** cells have a receptor for the α-factor, called Ste2, and the α cells have a receptor for the **a**-factor, called Ste3 (Fig. 2-1).

The Mating Process Involves Several Steps

When **a** and α cells encounter each other, they respond in several ways to pheromone produced by the other mating type (Fig. 2-1). In the initial response, cells stop dividing, transcription of a set of genes that encode proteins required for the mating process occurs, and cell growth becomes directed toward the mating partner. This polarized growth produces a structure called a mating projection, and a yeast cell with such a projection is called a "shmoo" (named after a creature in a comic strip). Because yeast cells have no means of propulsion, formation of a mating projection allows mating partners that are nearby, but not touching, to come into contact by growing their projections toward each other (Fig. 2-1).

Once the cell walls of the two mating partners make contact, they adhere to each other. Subsequently, this portion of the cell wall is degraded, allowing the underlying cell membranes to fuse to form a single larger cell with two nuclei (Fig. 2-1). The nuclei next migrate toward each other. The two nuclear membranes ultimately fuse to produce a single nucleus with a diploid complement of chromosomes (Fig. 2-1). At this point, an **a** and an α cell have mated to form an **a**/α cell.

a/α Cells Can Undergo Meiosis and Sporulation

a/α cells do not mate with each other nor with **a** or α cells. Yet, **a**/α cells can do several things that their haploid counterparts cannot. As mentioned in Chapter 1, diploid **a**/α cells can undergo the sequential processes of meiosis and spore formation. This occurs in response to the simultaneous starvation for carbon- and nitrogen-containing nutrients (see Fig. 1-1). The result of meiosis is the production of spores, which, under favorable environmental conditions, germinate to yield haploid mating-type **a** or mating-type α cells.

The processes of meiosis and spore formation have two functions. The first is to generate, under conditions of starvation, spores that can persist until the environment is more favorable. The second is to complete the sexual cycle of yeast to produce the haploid cells. The process of meiosis in all eukaryotes produces four haploid products (the details of how this occurs are beyond the scope of this book). Therefore, after an **a**/α cell undergoes meiosis and sporulation, four spores are produced and these are placed in a sac called an ascus (Fig. 2-2). Two of the spores

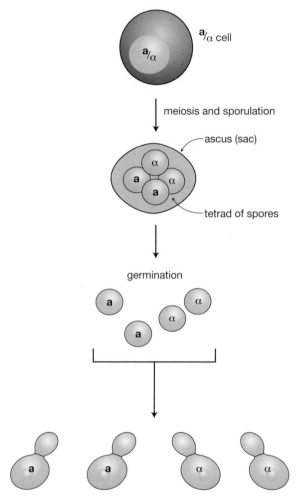

Figure 2-2. a/α diploid cells can undergo meiosis and sporulation to produce spores that can germinate into haploid cells.

will be of mating type **a**, whereas the other two will be of mating type α. Under favorable environmental conditions, the ascus is degraded, freeing the spores to convert into metabolically active, dividing yeast cells (Fig. 2-2).

a/α Cells Can Switch to a Filamentous Form of Growth upon Starvation for Nitrogen Nutrients

Under other environmental conditions, **a**/α cells can switch into a filamentous growth form rather than undergoing meiosis and sporulation: In response to starvation for a nitrogen-containing nutrient source, but given adequate carbon-con-

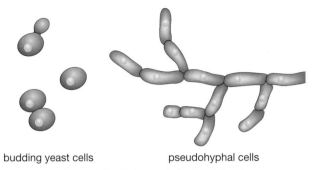

budding yeast cells   pseudohyphal cells

Figure 2-3. Yeast and pseudohyphal growth forms of *Saccharomyces cerevisiae*.

taining nutrients, **a**/α cells can switch from growing as oval-shaped cells to a filamentous form called pseudohyphae (Fig. 2-3). Pseudohyphal cells are elongated
and maintain attachments between the cell walls after division. When grown on a
semisolid medium, such as an agar plate, pseudohyphae can invade the medium.
Moreover, the cells secrete an enzyme that breaks down pectin, a major carbohydrate found in rotting fruit, where yeast normally lives. Given these properties of
pseudohyphal cells, it is thought that the ability of **a**/α cells to switch into this alternative growth mode confers the ability to forage for nutrients in the wild.

The **a** and α cell types do not have this response to nitrogen starvation, but do
switch to a filamentous mode of growth when starved for carbon-containing nutrients: This process has been termed "haploid filamentation" and is characterized by
formation of similar but shorter filaments. Why the environmental signals that trig-
ger filamentous growth differ for the haploid versus diploid cell types is unknown.

Because cells undergoing filamentous growth have shapes and properties that
distinguish them from nonfilamentous cells, they, like spores, can be considered
distinct cell types. Although spores are an intermediate stage between an **a**/α cell
type and haploid cell types, **a**, α, or **a**/α cells can switch into or out of these filamentous forms depending on signals from the environment.

Asexual Growth of Yeast Cells: The Cell Cycle

In addition to participating in mating and meiosis (the sexual cycle as described
above), the **a**, α, and **a**/α cell types can also proliferate asexually through cell division. Here, a yeast cell divides to produce two identical cells. This division occurs
by budding (Fig. 2-4). In this process, a cell grows an appendage—called a bud—
into which contents (such as a nucleus containing a replicated copy of the genome
as well as all of the other organelles) are placed prior to cell division. The cell that
is producing the bud is called a "mother" cell, whereas the bud is the incipient
"daughter" cell (Fig. 2-4).

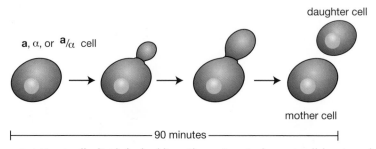

Figure 2-4. Yeast cells divide by budding. Shown is a single yeast cell forming a bud.

Yeast cells, like all cells, grow and divide through a sequence of events called the cell-division cycle or, more simply, the cell cycle. The cell cycle has two purposes with respect to the genome: (1) to replicate each chromosome and (2) to give each progeny cell one copy of each chromosome.

The overall cell cycle is divided into four sequential phases: G_1, S, G_2, and M. Figure 2-5 shows the states of the chromosomes during several of these phases, and Figure 2-6 shows views of yeast cells at each phase. Newly divided cells enter the G_1 phase, during which they grow to become significantly larger. Next, in S phase (which stands for synthesis of DNA), each chromosome is duplicated and the daughter bud begins to develop. After a brief G_2 phase, the cell enters the M phase (M stands for mitosis). During M phase, the two copies of each chromosome are segregated to opposite ends of the nucleus. Division of the nucleus and then the

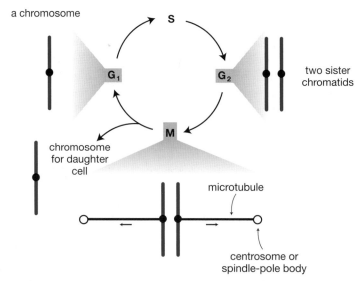

Figure 2-5. Chromosome behavior during the cell cycle. Microtubules and centrosomes are part of the apparatus that segregates chromosomes during mitosis.

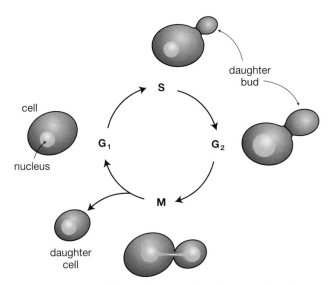

Figure 2-6. Morphology of yeast cells during the cell cycle.

cell results in two progeny cells, each of which possesses a copy of each chromosome. G_1 and G_2 are named as such because they are "gaps" in the cell cycle between the other two phases of the cell cycle (S and M).

The cell cycle can be regulated by signals from the environment. For example, when haploid cells encounter mating pheromone from the opposite mating type, they stop proliferating by arresting their cell cycle in the G_1 phase. After mating, the resulting **a**/α cell restarts its cell cycle in the G_1 phase. It is important that cells regulate the point in the cell cycle at which they mate. If two G_2-phase cells were to mate, for instance, then the resulting **a**/α cell would contain twice the normal number of chromosomes—four copies of each instead of two.

The Pattern of Budding Also Distinguishes the Cell Types

During "normal" cell division (i.e., outside of meiosis), the choice of the site at which the new daughter cell begins to form is not random (Fig. 2-7). Rather, cells choose their division site based on the location of the previous division site. This process is regulated differently in the haploid versus the diploid cell types. **a** and α cells form a bud adjacent to the previous site of budding. This pattern is called an "axial" pattern. **a**/α cells, in contrast, bud alternatively from both this side of the cell and the side opposite from the previous division site. This is called the "bipolar" pattern. Why these cell type–specific differences in the pattern of budding exist and how they occur will be covered in Chapter 8.

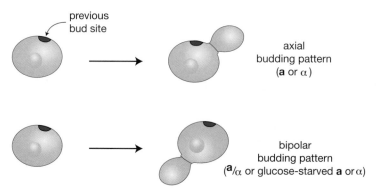

Figure 2-7. The site of cell division (budding) is chosen based on the location of the previous site of division. (Although not discussed in the text, cells undergoing haploid filamentation switch to a bipolar budding pattern in response to glucose starvation.)

Summary of the Properties of **a**, α, and **a**/α Cell Types

To summarize (Table 2-1), the **a**, α, and **a**/α cell types have distinctive properties. The haploid cell types are capable of mating, but only with the opposite mating type. **a**/α cells cannot mate, but, in response to environmental conditions, can undergo meiosis and sporulation or become nutrient-foraging pseudohyphal cells. Haploid cells can also switch to filamentous form. The budding pattern of yeast cells is nonrandom, and this pattern differs between haploid cell types and the **a**/α cell type. In the following chapters, we will come to understand the molecular mechanisms by which these differences among cell types occur.

Table 2-1. Some of the properties of yeast cells that are controlled by cell type

	Cell Type		
Property	**a**	α	**a**/α
Mating with **a**	no	yes	no
Mating with α	yes	no	no
Meiosis/sporulation	no	no	yes
Axial budding	yes	yes	no
Bipolar budding	no	no	yes
Pseudohyphal growth	no	no	yes
Haploid filamentation	yes	yes	no

MASTER REGULATORY TRANSCRIPTION FACTORS CONTROL CELL TYPE

In Chapter 1, I introduced the two haploid cell types of yeast, mating type **a** and mating type α. The third cell type is produced by mating of **a** and α cells, the diploid **a**/α type. We learned in Chapter 2 of two important differences between the **a** and α cell types: Each produces a distinct mating pheromone and each displays a receptor on its cell surface that specifically detects the pheromone produced by the opposite mating type. We also learned that the **a**/α cell type is incapable of mating, but is able to undergo meiosis and sporulation.

The Pattern of Gene Expression Distinguishes the **a**, α, and **a**/α Cell Types

These differences in behavior among the **a**, α, and **a**/α cell types are caused by different patterns of gene expression. For example, **a** cells—but not α cells—transcribe the genes encoding the **a**-specific pheromone (**a**-factor), and the receptor for the α-specific pheromone (the α-factor receptor Ste2). Likewise, α cells—but not **a** cells—transcribe the genes encoding α-pheromone and the **a**-pheromone receptor, Ste3. In **a**/α cells, which do not mate, neither the pheromones nor the receptors are produced because the genes that encode them are not transcribed.

Gene Regulation

Because the differences between **a**, α, and **a**/α cells involve the control of gene transcription, I will mention briefly some general features of gene regulation in eukaryotes that will be illustrated in this chapter.

Genes are controlled by regulators that bind to specific DNA sequences; in yeast, these sequences are usually found upstream of the coding sequences. A given regulator can either turn on (activate) or turn off (repress) the transcription of a gene to which its binds. As we will see in this chapter, regulators are often composed of several different polypeptide subunits.

The archetypal regulator has two domains: a DNA-binding domain that recognizes a specific sequence and a domain that is required for the regulator to control transcription once bound to DNA. Regulators that activate transcription ("activators") each contain a "transcriptional activation domain," whereas those that repress transcription ("repressors") each contain a "transcriptional repression domain." The function of activation and repression domains is to bring proteins to the promoter that more generally control transcription. Activation domains, for example, recruit protein complexes that in turn recruit RNA polymerase II, the enzyme responsible for synthesizing messenger RNA (mRNA) in eukaryotes.

With these principles in mind, we will spend the remainder of the chapter focusing in detail the mechanisms by which cell type–specific transcription of genes occurs.

The *MAT* Locus Is the Master Controller of Cell Type and Occurs in Two Versions

The distinct patterns of gene expression between **a** and α cells are determined by a single genetic locus, called *MAT* (Fig. 3-1). The *MAT* locus differs between the two cell types (Fig. 3-1) and is the only difference between the genomes of the two mating types. **a** cells have the *MAT***a** version of the mating-type locus, whereas α cells have the *MAT*α version.

Genes at the *MAT* locus encode proteins that bind specific DNA sequences. Two proteins, called **a**1 and **a**2, are encoded by the *MAT***a** locus (Fig. 3-1A). These are different from the two proteins encoded by the *MAT*α locus, which are named α1 and α2 (Fig. 3-1B). In **a**/α cells, both versions of *MAT* exist (Fig. 3-1C).

The Overall Scheme for Mating-type Regulation

The proteins encoded by the *MAT* locus associate with other proteins to form the regulators whose actions lead to the patterns of transcription that are characteristic of each cell type. Figure 3-2 provides a "cheat sheet" that summarizes the overall scheme of gene regulation in **a**, α, and **a**/α cell types. It may be useful for the reader to refer back to Figure 3-2 to assist in placing the individual pieces of information described below into the overall scheme.

A   Chromosome III

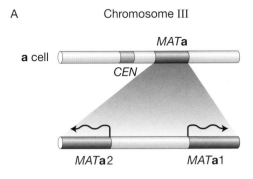

B

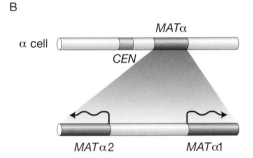

C

Figure 3-1. The mating-type locus, *MAT*.

α1 Is Required for the Activation of α-specific Genes

In this section, we will focus on genes that are expressed solely in α cells——α-specific genes (α-sgs). α1 binds directly to the promoters of α-sgs by recognizing a specific DNA sequence present in the promoters of this class of genes. In the absence of α1, α-sg transcription is not activated and the genes lie dormant. Since **a** cells do not contain the gene encoding α1, they cannot express α-sgs.

α-sgs include two redundant genes that encode α-factor, *MFα1* and *MFα2* (not to be confused with *MATα1* and *MATα2*!), as well as the **a**-factor receptor gene, *STE3* (Fig. 3-3).

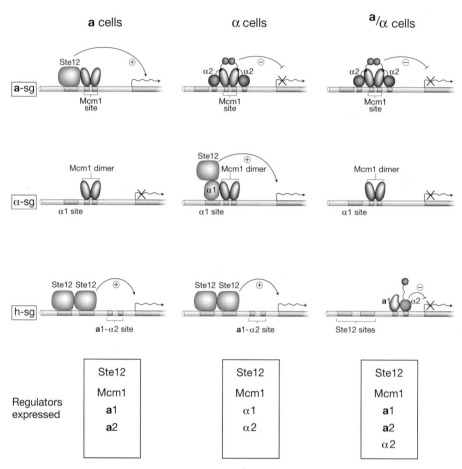

Figure 3-2. Overall scheme of cell type control. Shown is how **a**-specific genes, α-specific genes, and haploid-specific genes are regulated in **a**, α, and **a**/α cells. Also shown are the cell types in which the regulators are expressed. Note that one particular class of haploid-specific genes is shown—those that respond to the extracellular presence of mating pheromone. Also note that of the proteins shown, only Ste12 contains a domain capable of activating transcription.

Although the regulatory proteins encoded by the *MAT* locus are ultimately responsible for specifying cell type, they do not work alone. As shown in Figure 3-4, α1 binds to DNA along with two other proteins called Mcm1 and Ste12. These two proteins are expressed in all three cell types. The Mcm1 protein binds cooperatively as a dimer to a DNA site adjacent to the α1 binding site (Fig. 3-4). Cooperative binding is explained in Box 3-1. α1 and Mcm1 exhibit DNA binding cooperativity with each other as well.

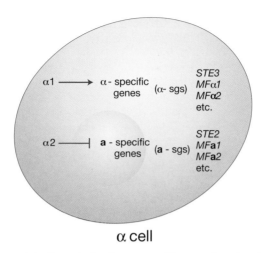

Figure 3-3. Control of cell type–specific genes by α1 and α2.

Binding to DNA and activation of transcription can be mediated by different polypeptides. For example, neither α1 nor Mcm1 are capable of activating transcription on their own—a third protein is required for α-sgs to be transcribed. This protein is Ste12, and among the three proteins specifically necessary for the expression of α-sgs, it is the only one that possesses a transcriptional activation domain. Ste12 does not contact the DNA directly at the promoters of α-sgs, but is recruited to the promoter by a protein–protein interaction with α1 (Fig. 3-4). Once recruited through this interaction, Ste12 activates transcription of α-sgs.

To summarize, α1 activates the transcription of α-sgs such as *STE3*. It does so by binding to DNA cooperatively with the Mcm1 protein and recruiting a third protein called Ste12.

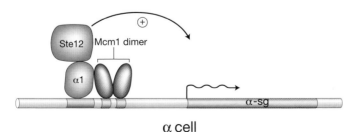

α cell

Figure 3-4. Activation of α-sgs by α1.

Cooperative binding of proteins to DNA is a common occurrence. By definition, it occurs when the presence of one DNA-binding protein lowers the concentration required by another protein to bind to DNA. The proteins can be two molecules of the same protein (as in the case of the cooperative binding of Mcm1 subunits to DNA) or different proteins (in the case of α1 binding cooperatively to DNA with Mcm1).

In its simplest manifestation, cooperative DNA binding between two proteins typically depends on the following features: (1) that the two proteins bind to sites that are linked to each other on the DNA, (2) that the two proteins can touch each other, and (3) that the concentrations of the proteins, and their affinities for DNA, are low enough that their binding to each other becomes necessary for the DNA to be occupied by one or both proteins.

What are the consequences of cooperative DNA binding?

One of them has been mentioned earlier in the chapter: Cooperativity allows for combinatorial control. What do I mean by this? By making the binding to DNA of one regulator depend, through cooperativity, on the binding of another, a given gene can be set to "switched on" (for example) only when both regulators are present. If each regulator is available to bind DNA only in response to a specific signal, then the gene is switched on only when both signals are present. This can be extended to more signals by making the binding of further regulators also depend on cooperativity.

By mixing and matching the DNA-binding sites for different regulators (which are able to bind cooperatively) within promoters of different genes, new combinations of signals can be required to switch on different genes, allowing a promoter to integrate signals.

Cooperative DNA binding can also be used to generate steep "all-or-none" effects. That is, the binding of a protein to DNA can be exquisitely sensitive to its concentration, with small changes in that concentration having dramatic effects on DNA site occupancy. Thus the state of gene expression, stable in one state, can be poised to completely switch to an alternative stable state over a very narrow change in regulator concentration. Although this property is not relevant to the discussion of the mating-type regulators described in this chapter, it is crucial for understanding gene regulation in other contexts, such as the genetic switch of phage λ.

Cooperativity also helps deal with an issue that arises from the fact that DNA-binding proteins not only bind to specific DNA sequences, but to other "nonspecific" DNA sequences, albeit with a lower affinity. This presents a problem for a given protein trying to find its site because the number of nonspecific sites in a genome is typically huge compared to the number of specific sites for that protein. Thus, even though the affinity of each nonspecific site is low—and thus each site holds the protein for only a very short time—the overall effect of the population of nonspecific sites can be immense. In effect, the protein may spend the vast majority of its time caught up in an endless sampling of the low-affinity sites.

Cooperativity overcomes this problem. Because of the large number of nonspecific sites, and because the protein samples each so fleetingly, it is unlikely that two molecules of the protein will simultaneously occupy adjacent nonspecific sites. Specific sites, with their higher affinity for the protein, hold that protein for longer, thus vastly increasing the chance that protein bound at one such site will make contact with another molecule of protein bound at an adjacent specific site (if such is available). The two proteins can then bind there cooperatively, stabilizing each other at those sites.

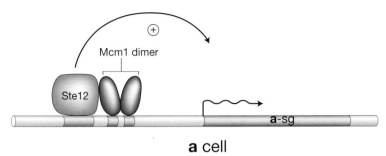

a cell

Figure 3-5. Activation of **a**-sgs by Ste12-Mcm1.

In **a** Cells, **a**-specific Genes Are Activated by a Ste12-Mcm1 Complex

In *MATa* cells, neither **a**1 nor **a**2 plays a role in the activation of **a**-specific genes. Rather, the activator protein Ste12 binds directly to DNA sites that exist in the promoters of **a**-sgs. These sites each occur adjacent to a binding site for the Mcm1 dimer, facilitating the cooperative binding of Ste12 and Mcm1 to the promoters of **a**-sgs (Fig. 3-5). This occurs in much the same way as α1 and Mcm1 bind to α-sg promoters in α cells, with the key difference being that **a**-sgs contain a DNA sequence recognized by Ste12 rather than the sequence recognized by α1 (Fig. 3-5).

From what we have learned so far, we can see that Ste12 can interact specifically with a number of different molecules. These include a specific DNA sequence present in the promoters of **a**-sgs, the proteins Mcm1 and α1, and molecules involved in the activation of transcription. The ability of Ste12 to bind to these different molecules is mediated by distinct segments of the protein.

α2 Is Part of a Repressor of **a**-specific Genes

If Ste12 and Mcm1 are expressed in both **a** and α cells, what prevents **a**-sgs from being expressed in α cells? The answer is that the α2 protein is a repressor that turns off **a**-sgs in α cells by binding to their promoters (Figs. 3-3 and 3-6). Like α1 and

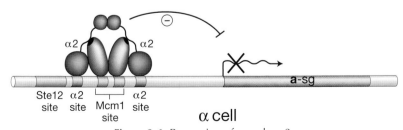

α cell

Figure 3-6. Repression of **a**-sgs by α2.

Ste12, α2 binds directly to DNA cooperatively with Mcm1, but it recognizes a different DNA sequence than either Ste12 or α1 (Fig. 3-6). Two molecules of α2 bind, one to either side of the Mcm1 dimer. They not only contact Mcm1, but also contact each other (Fig. 3-6). Thus, Mcm1 functions at both **a**-sgs and α-sgs, binding to different partners in the different cell types. Its primary function appears to be to provide a cooperative DNA-binding partner for α1, Ste12, or α2.

α2 Represses Transcription by Recruiting a General Corepressor

So how does α2 turn off **a**-sgs? A simple mechanism that one could imagine is that α2 occludes the site on Mcm1 that interacts with Ste12, thereby preventing Ste12 from binding to DNA cooperatively with Mcm1 (as implied in Fig. 3-6). Although this mechanism may play a role at some **a**-sgs, it does not fully explain how α2 acts to repress transcription. To accomplish repression, α2 brings additional proteins to the promoter. Specifically, it recruits a "corepressor complex" comprised of the proteins Ssn6 and Tup1 (Fig. 3-7). This complex causes the repression of transcription when brought to promoters that would otherwise be active. The Ssn6-Tup1 complex blocks the recruitment of RNA polymerase II to **a**-sgs in α cells, but the exact mechanisms by which it does so are not understood.

Tup1 is also involved in the repression of many other genes in yeast. For each class of genes, Tup1 is brought to DNA by a distinct DNA-binding repressor protein analogous to α2 (Fig. 3-8). Proteins related to Tup1 and involved in gene repression are found in multicellular organisms. For example, in the fruit fly, *Drosophila melanogaster,* a gene called *groucho* encodes a Tup1-like protein involved in regulating gene expression during development. As in yeast, it is recruited by a variety of DNA-binding repressor proteins to repress transcription. Flies lacking *groucho* die during embryogenesis and exhibit abnormal development of the nervous system.

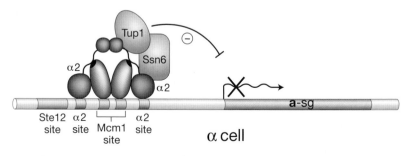

Figure 3-7. α2 represses genes by recruiting the Tup1-Ssn6 corepressor.

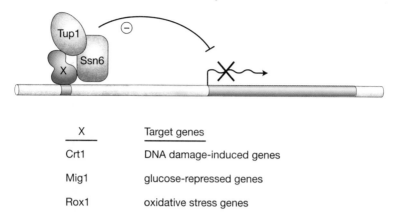

X	Target genes
Crt1	DNA damage-induced genes
Mig1	glucose-repressed genes
Rox1	oxidative stress genes

Figure 3-8. The Tup-Ssn6 corepressor is recruited to distinct gene sets by different DNA-binding proteins symbolized by "x." Each DNA-binding protein is inactivated by a particular environmental cue. For example, Crt1 normally represses a set of genes encoding proteins involved in the cellular response to DNA damage, and DNA damage causes the inactivation of Crt1, resulting in the transcription of these genes.

In **a**/α Cells, **a**1 (Encoded by *MAT***a**) and α2 Form a Repressor of Haploid-specific Genes

In this section, we will focus on the mechanisms that specify the diploid **a**/α cell type. As with the haploid **a** and α cell types, it is the pattern of gene expression that gives the **a**/α cell type its specialized characteristics.

As discussed in Chapter 2, **a**/α cells lack the ability to mate, but have the ability to undergo meiosis and sporulation. Because they are produced by mating between *MAT***a** and *MAT*α cells, **a**/α cells could in principle express all four genes encoded by the two mating-type loci. However, this is not the case. Instead, **a**/α cells express only three of the genes: α2, **a**1, and **a**2. Because α1 is not expressed, α-sgs remain dormant, and because α2 is expressed, **a**-sgs are also repressed.

How is α1 turned off? This is where the **a**1 protein comes in. Although it is expressed in **a** cells, **a**1 has no function in these cells. However, in **a**/α cells, **a**1 binds cooperatively with α2 to specific DNA sites that consist of a sequence recognized by **a**1 adjacent to a sequence recognized by α2. This **a**1-α2 complex functions as a repressor. It binds to a site in the promoter of the α1 gene, shutting off its transcription in **a**/α cells (Fig. 3-9A).

The **a**1-α2 repressor also binds to the promoters of another class of genes, the haploid-specific genes (h-sgs) (Fig. 3-9B). The h-sgs are defined as genes that are expressed in **a** cells and α cells, but not in **a**/α cells. Many h-sgs are involved in

A

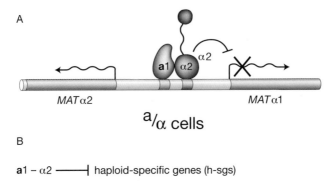

MATα2   MATα1

ᵃ/α cells

B

a1 – α2 ——| haploid-specific genes (h-sgs)

Figure 3-9. Repression of haploid-specific genes by **a**1-α2. (*A*) The **a**1-α2 complex represses the transcription of α1. (*B*) The **a**1-α2 complex represses the transcription of haploid-specific genes.

the mating responses of both **a** and α cells. These genes are activated by Ste12 bound to their promoters (see Fig. 3-2); their regulation will be covered further in Chapter 5. Because **a**/α cells express neither the pheromone nor the pheromone receptor genes (because of the presence of α2 and the absence of α1), they cannot mate. And even if these genes were to be expressed, mating would be blocked because of the repression of h-sgs by the **a**1-α2 repressor.

A Haploid-specific Regulator Explains Why **a** and α Cells Do Not Undergo Meiosis and Sporulation

A key h-sg is *RME1*, which stands for *Repressor of Meiosis 1* (Fig. 3-10). As its name suggests, expression of *RME1* in **a** and α cells prevents the expression of genes that are required for meiosis and sporulation. Because *RME1* is repressed by

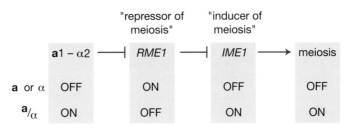

Figure 3-10. **a**1-α2 allows **a**/α cells to undergo meiosis by repressing the transcription of *RME1*, a repressor of meiosis. Shown are the expression states of *RME1* and its target gene *IME1* in **a** or α vs. **a**/α cells. Note that *IME1* expression requires that cells be starved for nitrogen- and carbon-containing nutrients.

a1-α2 in **a**/α cells, these genes are expressed and the diploid cell type can undergo meiosis and sporulation. So how does Rme1 work? Like α2 and **a**1-α2, Rme1 is a transcriptional repressor. It binds to the promoter of a gene called *IME1*, which stands for *Inducer of Meiosis 1* (Fig. 3-10). Ime1 is a transcriptional activator whose expression is sufficient to induce meiosis and sporulation. The promoter of the *IME1* gene is activated specifically under environmental conditions that induce meiosis and sporulation in yeast (a combination of nitrogen and carbon starvation). In haploid cells, the binding of Rme1 to the *IME1* promoter prevents *IME1* from being transcribed, even under conditions of carbon and nitrogen starvation.

The Second Protein Encoded by *MAT*a, a2, Has No Known Function

Up to this point, we have not discussed the second protein encoded by *MAT*a, a2. Remarkably, its function is still unknown. Cells of any type lacking the *MAT*a2 gene appear to behave normally. Thus, even in a well-studied model organism such as *Saccharomyces cerevisiae*, genetic mysteries remain.

The α2 and **a**1 Proteins Bind to DNA through Homeodomains

As described above, the control of cell type involves regulatory proteins that recognize specific DNA sequences. Building on earlier research on regulators in bacteria, studies in yeast were the first to identify such proteins in a eukaryotic organism. Insight into how a DNA sequence is recognized by yeast DNA-binding proteins has come from studies that have elucidated the atomic structures of proteins bound to DNA. The structure of the α2-Mcm1 complex bound to DNA and that of the **a**1-α2 complex bound to DNA have been determined (Fig. 3-11). They show that **a**1 and α2 bind to DNA through an evolutionarily conserved DNA-binding domain called the homeodomain (Fig. 3-12A). The term homeodomain comes from the fact that **a**1 and α2 are related to so-called "homeotic" genes that control the development the fruit fly (see Box 3-2). The homeodomain is an approximately 60-amino-acid segment consisting of three α helices that are packed against each other. The third helix is called the recognition helix and contains amino acids that contact the DNA directly and effectively "read" the sequence, in a manner similar to that seen in earlier studies of bacterial regulators (Fig. 3-12B). The protein segments that flank homeodomains are often involved in interactions with other proteins. For example, the region just downstream of the α2 homeodomain contacts **a**1 when the two proteins bind to DNA in **a**/α cells, and the sequence just upstream of the homeodomain contacts Mcm1 in α cells.

A

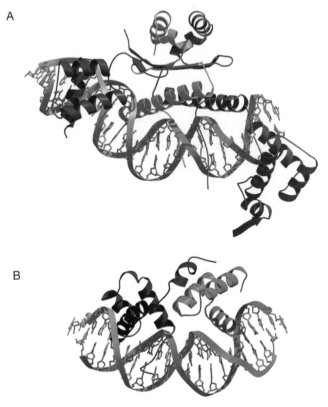

B

Figure 3-11. Structures of mating–type regulators bound to DNA. (A) α2-Mcm1-DNA structure (Tan and Richmond, *Nature 391:* 660 [1998]). Mcm1 dimer is in *blue* and *turquoise*, and the two α2 monomers are shown in *purple* and *pink*. Note that the portion of α2 that interacts with itself was not visualized in this crystal structure. (B) **a**1-α2-DNA structure (Li et al. *Science 270:* 262–269 [1995]). α2 is shown in *purple* and **a**1 is shown in *green*.

Mcm1 is part of an evolutionarily conserved group of proteins called MADS box proteins. Like the homeodomain, the MADS box is a domain that recognizes DNA via an α helix that makes specific contacts with the DNA. MADS is an acronym that stands for "*M*cm1, *A*gamous, *D*eficiens, and *S*RF (serum response factor)," which were the first four proteins identified with this domain. Agamous and Deficiens control flower development in plants, and SRF is a human transcription factor with roles in development and wound healing.

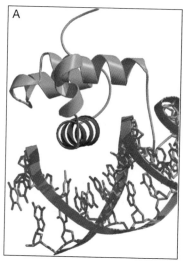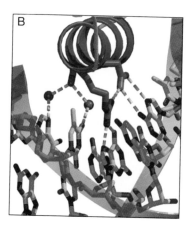

Figure 3-12. Homeodomain-DNA interactions (Li et al., *Science 270:* 262 [1995]). (*A*) α2 homeodomain bound to DNA. Note that the view is different from that shown in Fig. 3-11B. The recognition helix is indicated in *purple*. (*B*) Close-up of α2-DNA interactions showing hydrogen bonds (*dotted lines*) between the amino acid side chains of the recognition helix, water molecules (*red*), and bases in the DNA.

BOX 3-2. HOX GENES CONTROL PATTERNING DURING ANIMAL DEVELOPMENT

Homeobox, or Hox, genes encode homeodomain transcription factors. The investigation of genes that control the body pattern of *D. melanogaster* has revealed the involvement of many Hox genes. The Antennapedia gene of *Drosophila* is a Hox gene that initiates a cascade that results in the development of the animal's legs. A mutation that switches on this single gene in the antenna tissue of the fly produces legs in place of antennae, demonstrating how the expression of a single homeodomain protein can control development (Fig. 3-13). Many dozens of Hox genes control developmental patterns in *Drosophila*. Hox genes also control body pattern in mammals. The human genome encodes at least 42 families of Hox genes. Why the same DNA-binding domain should be used by diverse proteins to program development in organisms ranging from yeast to humans is not known. One possibility is that Hox genes evolved with the early transcriptional circuits that control development in the common ancestors of yeast and humans more than a billion years ago and that, although the circuits were modified extensively during evolution, the Hox genes themselves maintained their fundamental role within those circuits. Such a view would imply that the basic circuits that control cell type evolved only once in the distant past and were modified through mutation and natural selection through eons of evolution.

(Continued)

BOX 3-2. (*CONTINUED*)

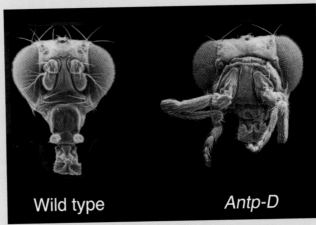

Figure 3-13. Antennapedia. Shown are scanning electron micrographs of the heads of *Drosophila* that illustrate the antenna-to-leg transformation caused by the *Antp-D* mutation.

Review Articles and Book

Johnson A.D. 1995. Molecular mechanisms of cell-type determination in budding yeast. *Curr. Opin. Genet. Dev.* **5:** 552–558.

Ptashne M. 2004. *A genetic switch: Phage lambda revisite*d, 3rd ed. Cold Spring Harbor Laboratory Press, Cold Spring Harbor, New York.

Sprague G.R. 2005. Three pronged genomic analysis reveals yeast cell-type regulation circuitry. *Proc. Natl. Acad. Sci.* **102:** 959–960.

CHAPTER FOUR

MAKING AND SECRETING CELL TYPE–SPECIFIC SIGNALS

Eukaryotic cells communicate with each other by secreting and responding to a variety of molecules. Unicellular organisms such as yeast secrete signals to facilitate a number of processes including mating. Cells in complex multicellular organisms secrete locally acting molecules involved in the maintenance of particular tissues as well as molecules called hormones that travel distantly to alter the physiology or behavior of the entire organism. These signals can be short unstructured peptides, larger folded proteins, or even lipids, and they can be made and released from cells in a number of different ways. This diversity illustrates a general principle in evolution—namely, that there is more than one than one way in which cells can evolve a mechanism that accomplishes a particular goal.

This chapter focuses on how the two yeast mating pheromones are synthesized and secreted. As we shall see, α-factor and **a**-factor differ in their modes of synthesis and secretion, providing an example of the principle outlined above. The mechanisms in yeast are similar to those used by human cells to synthesize and secrete peptide hormones.

α-Factor Is a Peptide Produced by Proteolytic Processing of a Larger Precursor

As we saw in Chapter 3, the mating pheromone produced by α cells—α-factor— is coded by the *MFα1* and *MFα2* genes. These genes are transcribed only in α cells, and their mRNA is translated to produce a polypeptide called pre-pro-α-factor. The precursor protein encoded by the *MFα1* gene contains four copies of the mature factor's amino acid sequence (Trp-His-Trp-Leu-Gln-Leu-Lys-Pro-Gly-Gln-Pro-Met-Tyr) embedded within a larger protein (Fig. 4-1). (That encoded by the

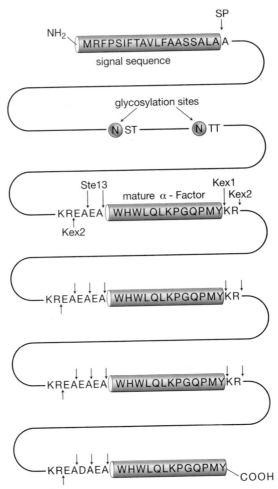

Figure 4-1. Structure of pre-pro-α-factor. Mature pheromone is indicated as cylinders. Cleavage sites for the proteases signal peptidase (SP), Kex2, Kex1, and Ste13 are indicated. Note that Kex2 must cleave at its sites before Kex1 or Ste13 can act. This is because Ste13 cleaves only downstream of sequences D-A or E-A when present at the amino terminus of a protein, and Kex1 cleaves only upstream of dibasic residues when they are at the carboxyl terminus of a protein. The precursor shown is encoded by the *MFα1* gene.

MFα2 gene is similar but contains six repeats of the sequence and will not be discussed further.) Thus, the pre-pro-α-factor must be cleaved to produce the mature pheromone. In the next two sections, we will learn how pre-pro-α-factor is processed and how it is released from the cell.

After Its Synthesis pre-pro-α-Factor Is Translocated into the Lumen of the Endoplasmic Reticulum

In eukaryotes, proteins destined for secretion from the cell are usually transported sequentially through two membrane-enclosed intracellular organelles called the endoplasmic reticulum (ER) and the Golgi apparatus. Proteins are targeted for secretion by a short peptide sequence at their amino termini called the "signal sequence." Signal sequence–bearing proteins are translocated into the ER and then transported in small membranous vesicles to the Golgi apparatus before being delivered (again in vesicles) to the plasma membrane of the cell. During their transit through these organelles, proteins are usually modified and processed in a number of ways.

The "pre" portion of pre-pro-α-factor is a signal sequence that targets it for translocation into the ER. This amino acid sequence is recognized by a complex of proteins in the membrane of the ER, the Sec61 complex (Fig. 4-2). The Sec61 complex is conserved across evolution (from bacteria to humans) and in eukaryotes it forms a pore through which virtually all secreted proteins are translocated into the lumen of the ER as the first step in their secretion. The high-resolution structure of the complex has been solved (Fig. 4-3) and displays a central pore that is normally plugged by a segment of the Sec61 polypeptide. It is thought that the signal sequence binds to the Sec61 complex in a way that displaces the plug, allowing the rest of the pheromone precursor to be threaded through the pore (in an unfold-

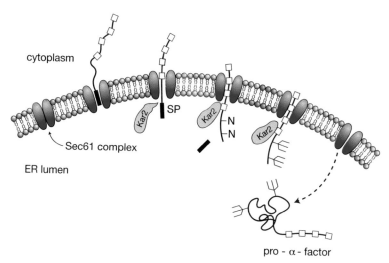

Figure 4-2. Translocation of pre-pro-α-factor into the lumen of the endoplasmic reticulum (ER).

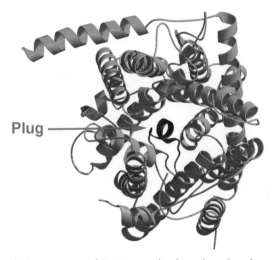

Figure 4-3. High-resolution structure of Sec61 complex from the achaeal organism *Methanococcus jannaschii* (van den Berg and Rapoport, *Nature 427:* 36 [2004]). This view is perpendicular to the plane of the membrane.

ed form). This process requires energy, which is provided by several proteins including Kar2, an abundant protein in the lumen of the ER that is capable of using the energy of ATP hydrolysis to help "pull" proteins into the ER (Fig. 4-2).

As the pre-pro-α-factor is translocated into the ER lumen, its signal sequence is removed by a protease called signal peptidase (called "SP" in Fig. 4-2), which is found in the ER lumen. The pro-α-factor is then modified on specific asparagine residues by the addition of sugars, a process called glycosylation (indicated by "pitchforks" in Fig. 4-2). Glycosylation is important for the transport of pro-α-factor to the Golgi apparatus via membrane vesicles.

Pro-α-Factor Is Cleaved to the Mature Form in the Golgi Apparatus and Then Sent to the Plasma Membrane for Secretion from the Cell

On arrival in the Golgi apparatus, the α-factor precursor is cleaved by the proteases called Ste13, Kex1, and Kex2 to yield the mature α-factor peptide (Fig. 4-4; see also Fig. 4-1 for the locations of the cleavage sites). After processing, the mature α-factor is transported to the plasma membrane in vesicles. Secretion occurs when the vesicles fuse with the cellular membrane, thereby releasing α-factor into the extracellular space (Fig. 4-4).

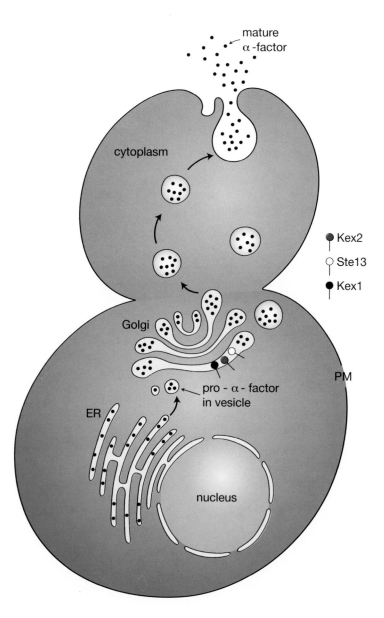

Figure 4-4. Transport of α-factor and its precursors through the secretory system. Dots within the ER, vesicles, and Golgi apparatus denote molecules of pro-α-factor.

BOX 4-1. POMC IS A HUMAN HORMONE PRECURSOR THAT, LIKE PRO-α-FACTOR, IS CLEAVED BY KEX2-LIKE PROTEASES TO PRODUCE MATURE HORMONES

Many hormones in the human body are cleaved from a large precursor in the Golgi apparatus. One dramatic example is pro-opiomelanocortin (POMC), which is cleaved to produce four different hormones: melanocyte-stimulating hormone, adrenocorticotropin (ACTH), lipotrophin, and endorphin (Fig. 4-5A). These hormones have many important effects on the body. For example, ACTH stimulates release of the steroid hormone cortisol from the adrenal gland, which has important anti-inflammatory properties. POMC is made in the largest amounts in cells of the anterior pituitary gland, which sits at the base of the brain. Once POMC is cleaved into its constituent hormones, they are released into the bloodstream. Proteases with a high degree of similarity to Kex2 catalyze the cleavage of POMC downstream of dibasic residues (Lys-Lys, Arg-Arg, Lys-Arg, Arg-Lys). The high degree of conservation of this mechanism is exemplified by the fact that the yeast Kex2 protein can substitute for the human homolog in the cleavage of POMC and the consequent production of hormones. Indeed, the identification of Kex2 in yeast was the key breakthrough that permitted the identification of the human enzymes. The three-dimensional structures of Kex2 and one of its human homologs, called furin, have been solved. Given the high degree of functional conservation, it is not surprising that the two structures are virtually identical (Fig. 4-5B).

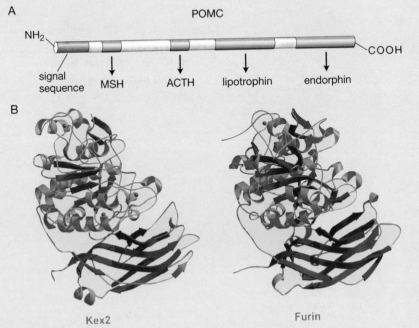

Figure 4-5. Prohormone processing. (A) The structure of pro-opiomelanocortin (POMC), a human prohormone. Three different hormones are produced from this precursor by proteolytic cleavages. (B) Crystal structures of the yeast Kex2 and human furin (*left*, Holyoak et al., *Biochemistry 42*: 6709 [2003]; *right*, Henrich et al., *Nat. Struct. Biol. 10*: 520 [2003]).

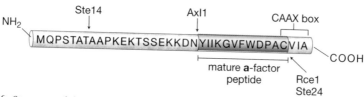

Figure 4-6. Structure of the **a**-factor precursor. The names of the proteases involved in the processing of **a**-factor are shown. Rce1 and Ste24 are so-called "CAAX proteases" that cleave downstream of cysteine residue of the CAAX box.

a-Factor Is Modified by the Addition of a Methyl Group and a Lipid in the Cytoplasm

Like α-factor, the **a**-factor peptide is also initially synthesized as a precursor polypeptide. However, at this point, the similarities end. Because it does not contain a signal sequence, the **a**-factor precursor remains in the cytoplasm where it is processed. It contains only a single copy of the mature peptide sequence near its carboxyl terminus (Fig. 4-6). The precursor ends with the sequence Cys-Val-Ile-Ala, which corresponds to the Cys-aliphatic-aliphatic-X-COOH motif, or so-called CAAX box. This sequence targets the protein for covalent lipid modification. The first step in this process is cleavage, which occurs immediately downstream of the CAAX box cysteine by means of the so-called CAAX proteases (see Fig. 4-6). A complex of two proteins then catalyzes the addition of the lipid prenyl group (Fig. 4-7). Following prenylation, the carboxyl group of the cysteine residue is methylated. Two additional proteases then act in sequence to produce the mature **a**-factor peptide (Figs. 4-6 and 4-7). All of the modifications of **a**-factor (proteolytic cleavages, prenylation, and methylation) turn out to be necessary for the pheromone to bind the Ste3 receptor on α cells. It is not obvious why **a**-factor must be so extensively modified in order to function properly, because α-factor requires no processing other than proteolytic cleavage.

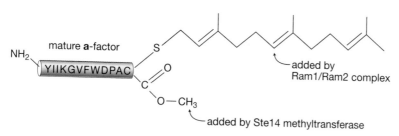

Figure 4-7. Structure of mature **a**-factor.

Box 4-2. Roles of Human Homologs of Enzymes That Process the a-Factor Precursor

The protease that produces the amino terminus of the mature **a**-factor, Axl1, is related to a human protein that is involved in cleaving the hormone insulin from its precursor in the islet cells of the pancreas. In addition, enzymes mediating the carboxy-terminal modifications of **a**-factor are also conserved (and it was the identification of the enzymes in yeast that led to the discovery of homologous enzymes in humans). In both yeast and humans, many proteins are prenylated in response to the CAAX signal at their carboxyl termini. Humans have a homolog of one of the yeast CAAX proteases that cleaves a protein called Ras that is involved in the genesis of human cancers. As with α-factor, cleavage downstream of the CAAX box cysteine is required for the attachment of a lipid. Unlike **a**-factor, Ras is not secreted; rather, Ras requires the attachment of the lipid to direct it to the plasma membrane, where it acts as an intracellular signaling molecule. Because mutations that activate Ras inappropriately are commonly found in human cancers, there has been great interest in inhibiting its processing.

a-Factor Is Secreted by a Specific Membrane Exporter Related to the Human Protein That Is Defective in Cystic Fibrosis Patients

In the previous section, we learned that that the processing of **a**-factor is quite different from that of α-factor. The mechanism by which **a**-factor is secreted also differs from that of α-factor. Unlike pre-pro-α-factor, the precursor to **a**-factor lacks a signal sequence and is therefore not translocated into the ER lumen. How, then, is **a**-factor secreted from the cell? The answer is that a protein in the plasma membrane, called Ste6, transports **a**-factor out of the cell (Fig. 4-8). This method of protein secretion is unusual, but may be necessary for lipid-modified peptides that would otherwise embed themselves into membranes. *STE6* is itself an **a**-specific gene. Its only function is to secrete **a**-factor, and it is therefore not needed in α cells. Ste6 is a member of a large class of membrane proteins involved in transporting molecules called ABC proteins. ABC stands for "ATP-binding cassette."

Ste6 membrane topology

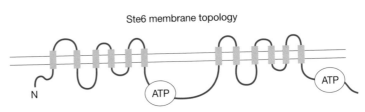

Figure 4-8. Membrane topology of the **a**-factor exporter Ste6.

Box 4-3. Cystic Fibrosis Is Caused by Mutations in an ABC Protein

A notable member of the ABC (ATP-binding cassette) family is the human protein CFTR (cystic fibrosis transmembrane conductance regulator) (Fig. 4-9). CFTR is a channel that conducts anions such as chloride. ATP binding regulates the activity of the channel—it opens the channel. (In this respect CFTR differs from members of the ABC family, such as Ste6, that couple the energy of ATP hydrolysis to a transport event.) Loss of CFTR causes the common and deadly inherited autosomal human disease cystic fibrosis (CF). Approximately 1 in 3000 Caucasians are born with CF, which is a recessive genetic disorder. CF patients suffer from severe lung infections and defects in the secretion of enzymes by the pancreas. The abnormalities in the lung are caused by a defect in the CFTR-mediated secretion of anions from the cells that line the airways. As a result of this defect, the mucus in the lung becomes thick and difficult to clear, thereby creating an environment in which bacteria thrive.

CFTR membrane topology

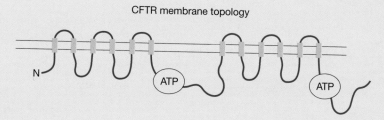

Figure 4-9. Membrane topology of CFTR, the protein defective in the inherited disease cystic fibrosis.

Members of this group have one or more ATP-binding domains that are usually required for their ability to transport molecules. For example, the binding and hydrolysis of ATP by Ste6 is coupled to its transport of **a**-factor.

Review Articles

Ashby M.N. 1998. CaaX converting enzymes. *Curr. Opin. Lipidol.* **9:** 99–102.

Maurer-Stroh S., Washietl S., and Eisenhaber F. 2003. Protein prenyltransferases. *Genome Biol.* **4:** 212.

Osborne A.R., Rapoport T.A., and van den Berg B. 2005. Protein translocation by the Sec61/SecY channel. *Annu. Rev. Cell Dev. Biol.* **21:** 529–550.

Rockwell N.C. and Thorner J.W. 2004. The kindest cuts of all: Crystal structures of Kex2 and furin reveal secrets of precursor processing. *Trends Biochem. Sci.* **29:** 80–87.

CHAPTER FIVE

DETECTING AND RESPONDING TO A SIGNAL

As described in Chapter 2, mating involves many steps, including the activation of gene expression, the arrest of the cell cycle, polarized growth of the shmoo toward the mating partner, cell–cell fusion, and nuclear fusion. This chapter focuses on the questions of how the binding of pheromone to its receptor on the cell surface activates gene transcription, how it causes the cell cycle to arrest in the G_1 phase, and how it causes the formation of a shmoo. The molecular chain of events that underlies these processes is generically referred to as "signaling" or "signal transduction" because it occurs in response to a particular extracellular signal—in this case, molecules of mating pheromone. Except for the pheromone receptors themselves, the molecules that mediate signaling are the same in **a** and α cells.

Signal Transduction

Signaling pathways convey information from the outside of the cell to the inside. In the case of cells responding to mating pheromone, this information includes whether mating pheromone is present, its concentration, and how long it has been present. This information is assessed continuously over time, so that if a signal disappears, processes that were initiated in response to that signal are terminated. Because of these requirements, proteins involved in signaling generally have the ability to be switched on and off rapidly. Whether a protein is in the "on" or "off" state is most often determined by a reversible modification such as protein phosphorylation (catalyzed by enzymes called "protein kinases") or by binding to a small molecule that can be hydrolyzed such as GTP. Typically, more than one protein of this type act in a sequence in response to a given signal. Having multiple components that can switch on and off in a pathway affords multiple points of control and the opportunity to have "forks" in the pathway where one protein can control two different processes. Each of these principles will be illustrated in this chapter.

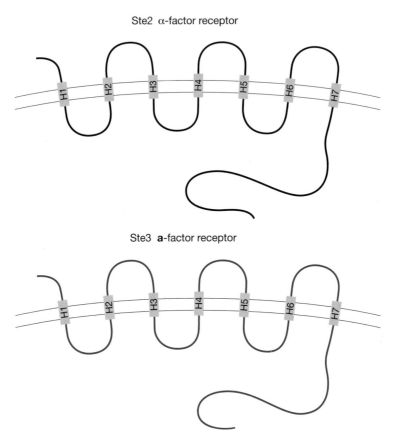

Figure 5-1. Membrane topologies of the pheromone receptors. The seven α-helices that span the plasma membrane are indicated as H1–H7.

Pheromone Receptors Are Seven-Transmembrane Proteins That Are Coupled to a Heterotrimeric G Protein

As we learned in Chapter 3, **a** cells can respond to α-factor because they transcribe the gene encoding the α-factor receptor, *STE2*. In contrast, α cells transcribe the gene for the **a**-factor receptor, *STE3*. Both Ste2 and Ste3 are members of a highly conserved family of cell surface receptors. These are the so-called seven-trans-membrane receptors, also known as G protein–coupled receptors (GPCRs; Fig. 5-1 shows the membrane topology of Ste2 and Ste3), for reasons that will become clear below. In humans, GPCRs mediate vision, smell, and many other responses; this evolutionary conservation underscores the ancient origins of this important family of "detector" proteins. Figure 5-2 shows the high-resolution structure of a typical GPCR. In most GPCRs, the binding of a ligand to the external surface of the

Figure 5-2. Crystal structure of rhodopsin, a G protein–coupled receptor (Palczewski et al., *Science 289:* 739 [2000]). Rhodopsin is the protein in the rod cells in the retina of the human eye that responds to light. The seven α-helices that span the membrane are indicated in color. Note that segments that were not ordered in the crystal appear as gaps in the peptide chain.

protein causes a change in the conformation of the receptor. For both Ste2 and Ste3, the binding to one of the mating pheromones controls their ability to interact with an intracellular protein complex called a heterotrimeric (having three subunits) G protein. In human cells there are a variety of such G protein complexes, but in *Saccharomyces cerevisiae*, a single complex exists, which mediates signaling from either the Ste2 or Ste3 receptor. This G protein is made up of subunits called Gpa1, Ste4, and Ste18 (Fig. 5-3).

The three subunits of G proteins are generically referred to as α (which binds GTP or GDP—this is where the "G" in "G proteins" comes from), β, and γ. Gpa1 is a G_α subunit, Ste4 is a G_β subunit, and Ste18 is a G_γ subunit. For simplicity, I will refer to Gpa1, Ste4, and Ste18 as G_α, G_β, and G_γ, respectively, for the remainder of the chapter. The G_β and G_γ subunits form a complex that binds to G_α when the latter is bound to GDP. Binding of pheromone to its receptor causes the receptor to interact with G_α, which in turn causes its bound GDP to be released and exchanged for GTP (Fig. 5-3B). Once this happens, G_α loses its ability to bind to the $G_{\beta\gamma}$ complex (Fig. 5-3B). The liberated $G_{\beta\gamma}$ complex is then capable of interacting with proteins that it could not when bound to G_α (Fig. 5-3B). Over time, GTP bound to G_α is hydrolyzed to GDP by G_α itself. Thus, if pheromone is no

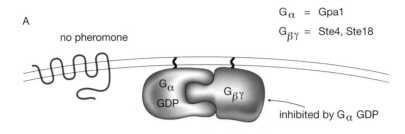

A

no pheromone

G_α = Gpa1

$G_{\beta\gamma}$ = Ste4, Ste18

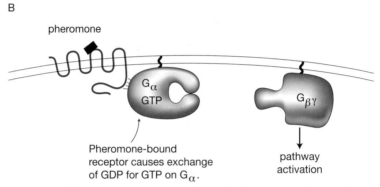

B

pheromone

Pheromone-bound receptor causes exchange of GDP for GTP on G_α.

pathway activation

Figure 5-3. Binding of pheromone to the receptor controls the G protein by stimulating exchange of GDP for GTP on G_α. The receptor binds to G_α only in the presence of pheromone. This binding induces exchange of GTP for GDP. When bound to GTP, G_α can no longer bind $G_{\beta\gamma}$. The free $G_{\beta\gamma}$ complex activates the remainder of the pathway. Note that the G protein subunits are covalently attached to lipid molecules that tether them to the plasma membrane.

longer bound to its receptor, G_α reverts to its original state and then binds and inhibits the $G_{\beta\gamma}$ complex. As described above, this ability to cycle between "on" and "off" states is a common feature of signaling proteins.

To summarize, the first step in pheromone signaling is the binding of pheromone to its receptor, which then causes the release of $G_{\beta\gamma}$ from its G_α inhibitor.

BOX 5-1. A BRIEF WORD ABOUT PROTEIN PHOSPHORYLATION

In this chapter, I will discuss many different protein phosphorylation reactions. As shown in Figure 5-4A, this process requires that a protein kinase bind ATP and its substrate and then transfer a phosphate from ATP to a hydroxyl group on the protein (recall that hydroxyl groups are found on the amino acids serine, threonine, and tyrosine). After the transfer, the substrate and the remaining ADP then dissociate from the kinase. For brevity, I will use two types of shorthand to describe phosphorylation reactions as shown in Figure 5-4B.

(Continued)

BOX 5-1. (*CONTINUED*)

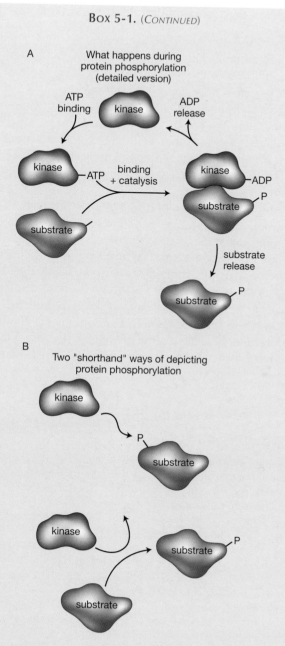

Figure 5-4. Protein phosphorylation. "-P" indicates phosphates covalently attached to proteins, whereas "-ATP" and "-ADP" represent noncovalent binding of ATP and ADP, respectively.

The Liberated $G_{\beta\gamma}$ Complex Protein Turns on a Kinase Cascade

In the pheromone response, the free $G_{\beta\gamma}$ complex activates a pathway in which four protein kinases are linked in series to form a "kinase cascade." Each kinase in the pathway phosphorylates the next kinase, converting the next kinase from its inactive form to its active form (Fig. 5-5). Activation of final kinase in this relay mechanism triggers both the arrest of the cell cycle and the transcription of genes involved in mating.

The $G_{\beta\gamma}$ complex recruits a protein called Ste5 to the plasma membrane (Fig. 5-5). Ste5 in turn brings with it three protein kinases called Ste11, Ste7, and Fus3. Because of its ability to bind multiple signaling proteins, Ste5 is referred to as a "scaffold" protein.

Ste11, Ste7, and Fus3 form part of the kinase cascade (Fig. 5-5). That is, Ste11, when active, phosphorylates Ste7 on two amino acid residues—thereby converting Ste7 from an inactive conformation to an active structure. In turn, active Ste7 phosphorylates Fus3 on two specific sites, resulting in the activation of its kinase activity. Fus3 then leaves the scaffold at the plasma membrane and enters the nucleus. How does the recruitment of the scaffold–kinase complex to the plasma membrane result in the activation of Ste11? Like Fus3 and Ste7, Ste11 must be phosphorylated to be an active kinase. This is done by yet another kinase called Ste20, which is active only at the plasma membrane because it is bound and activated by a membrane-associated small monomeric GTPase called Cdc42 (more on

---

BOX 5-2. THE HUMAN MAP KINASE CASCADE IS ACTIVATED IN CANCER CELLS

The system of kinases described in this chapter is called a MAP kinase cascade because homology to kinases in human cells was discovered during the study of mitogens, molecules that stimulate cell division. The term "MAP kinase" stands for *mitogen-activated protein kinase*. (These are also sometimes referred to as ERK for *extracellular signal-regulated kinase*.) Fus3 is the MAP kinase in the yeast pheromone response pathway. Ste7 is homologous to a human protein called MAP/ERK kinase (MEK) because it phosphorylates MAP kinase. Likewise, Ste11 is homologous to MAP/ERK kinase kinase (MEKK), which was named as such because it phosphorylates MEK. Finally, moving up to the top kinase of the hierarchy, Ste20 is called p21-activated kinase (PAK) because the molecular weight of its activator Cdc42 is 21 kD. These names are indicated next to the homologous yeast kinases in Figure 5-5. In human tumors, mutations in the *ras* oncogene described in Chapter 4 cause unregulated activation of the MAP kinase cascade and uncontrolled cell proliferation. Activating mutations in *ras* occur in at least 30% of human cancers and as a consequence those tumors have the MAP kinase cascade and cell proliferation program turned on whether or not extracellular growth control signals are present.

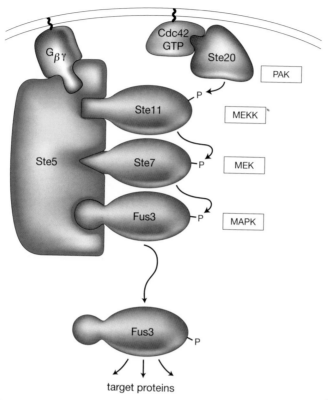

Figure 5-5. $G_{\beta\gamma}$ activates the MAP kinase signaling cascade by recruiting the scaffold protein Ste5 to the plasma membrane. Note that Cdc42 is covalently modified with a lipid molecule that tethers it to the membrane. The names of the corresponding protein kinases in human cells are shown in boxes.

how Cdc42 is itself activated in a moment). The function of the scaffold protein Ste5 is to tether the kinase together so that one kinase can phosphorylate the next.

For any phosphorylated protein, there exist enzymes called protein phosphatases that remove the phosphates. For the most part, these enzymes are not regulated and thus continuously dephosphorylate proteins in the cell. Moreover, phosphatases usually are not highly specific enzymes, and often several different phosphatases are capable of removing a particular phosphate from a protein. The importance of protein dephosphorylation for the pheromone response mechanism is that each kinase in the cascade requires the continuous presence of the active, upstream kinase in order to remain active. If cells no longer are exposed to mating pheromone, the entire cascade will rapidly shut down as a result of the action of phosphatases.

Why a Cascade?

What is the purpose of having a cascade of kinases? Would it not be simpler to have fewer components? It has been argued that the cascade allows for amplification of the input signal. This is because each kinase is an enzyme and can therefore phosphorylate many molecules of the downstream kinase. However, there cannot be a great deal of signal amplification in the pheromone signaling pathway: Fus3 is present at only about ten times the concentration of its activator Ste7, whereas Ste11 and Ste20 are present at roughly the same levels as Ste7.

Another proposal for why the cascade design is used comes from computer modeling studies. These suggest that the combination of reversible protein phosphorylation and the cascade can lead to very sharp "all or none" behavior. By this, I mean that below a certain threshold concentration of the signal, there is no signaling, but above that threshold, the pathway becomes fully activated. Although this behavior has been observed in kinase cascades in some organisms, the pheromone response pathway does not exhibit this behavior. Rather, the activity of the pathway appears to be "graded" such that increases in pheromone concentration to which cells are exposed lead to increases in the activity of the kinases of the pathway over a large range.

A more likely general explanation for the use of a kinase cascade has been alluded to above: Having multiple components in a signaling pathway allows for more opportunities for the pathway to be regulated. It may be that having multiple "control nodes" in the pathway allows it to relay information in a more precise manner.

Pheromone Signaling Leads to the Arrest of the Cell Proliferation through the Activation of an Inhibitor of the Cell Cycle

As described in Chapter 2, **a** and α cells respond to the appropriate pheromones by arresting the cell cycle. The key protein involved in this response is Far1 (for "feromone arrest" [sic]). When phosphorylated on a specific site, the Far1 protein is capable of blocking the cell cycle in the G_1 phase. This site can be phosphorylated by only one protein kinase, namely Fus3, the terminal kinase in the cascade described above. Thus, exposure of cells to pheromone produced by the opposite mating type activates Fus3. Active Fus3 phosphorylates Far1, and this causes cells to arrest their cell cycle in preparation for mating.

The Transcriptional Response to Pheromone Integrates Cell Type and Environmental Signals

As mentioned in Chapter 3, exposure of cells to pheromone results in the transcription of numerous genes involved in mating. Among these are the genes for the pheromone receptors and for the pheromones themselves. Thus, once cells initial-

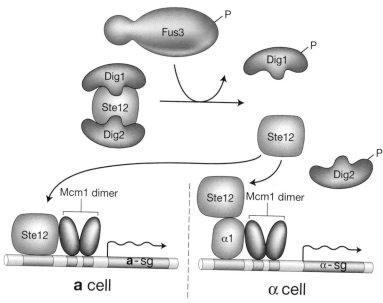

Figure 5-6. The Fus3 MAP kinase activates the transcription factor Ste12 by phosphorylating and inactivating inhibitors of Ste12.

ly detect pheromone from the opposite mating type, they increase their capacity to respond to and produce mating pheromone. Presumably this boost in the expression of the mating machinery ("getting excited") increases the chances of successful mating.

How does activation of Fus3 result in increased transcription of these and other genes? The answer is that Fus3 activates the transcription factor Ste12 that we encountered in Chapter 3 (Fig. 5-6). Recall that Ste12 promotes the expression of **a**-specific genes and α-specific genes. In the absence of signaling, Ste12 is inhibited by the proteins Dig1 and Dig2 that each bind to different parts of the Ste12 molecule—Dig2 to the DNA-binding domain of Ste12 and Dig1 to a region near to carboxyl terminus that is required for DNA-bound Ste12 to activate transcription. Upon their phosphorylation by active Fus3, Dig1 and Dig2 release Ste12, allowing it to bind to the promoters of **a**-sgs or α-sgs (depending on the cell type) and activate their transcription. The ability of Ste12 to activate the promoters of **a**-sgs and α-sgs provides a mechanism for cells to connect their cell type (as determined by the transcriptional regulators encoded by the *MAT* locus, as we discussed in Chapter 3) with environmental information (the presence of one or the other mating pheromone).

As mentioned in Chapter 3, Ste12 also activates a subset of haploid-specific genes (h-sgs) that are involved in mating. At these h-sgs, Ste12 does not bind in cooperation with Mcm1 or α1 as it does at cell type–specific promoters. Instead, it binds as a homodimer (see Fig. 3-1). Binding to DNA by itself (rather than with a cell type–specific partner such as α1) to this class of genes makes sense because h-sgs by definition are active in **a** cells as well as α cells.

If Ste12 requires pheromone signaling to be active, how can it mediate the transcription of **a**-sgs, α-sgs, and h-sgs in the absence of pheromone? It turns out that a low level of signaling (often termed "basal signaling") is maintained throughout the pheromone response pathway, even when pheromone is absent, and this level is sufficient to activate enough Ste12 to maintain a baseline of transcription. The mechanism by which basal signaling occurs is unclear. This low level of activation is not sufficient to activate enough Far1 to cause arrest of the cell cycle.

The Activated G Protein Directs Mating Projection Formation toward the Mating Partner by Recruiting Cell Polarity Factors That Promote Actin Polymerization

As described in Chapter 2, a notable feature of mating is that cells shmoo toward their mating partners. Here is how it is thought to work (Fig. 5-7): The pheromone receptors most activated on the cell surface are those facing the direction containing the highest concentration of pheromone. As a consequence, this region of the cell has the highest concentration of active $G_{\beta\gamma}$, and $G_{\beta\gamma}$ recruits proteins that promote polarized cell growth in that direction.

Which proteins does the $G_{\beta\gamma}$ complex recruit to promote polarized cell growth? Surprisingly, the key link is the Far1 protein, which turns out to have a second role in mating that is unrelated to its role as a cell cycle inhibitor. Far1 binds to the free $G_{\beta\gamma}$ complex (Fig. 5-7). In turn, Far1 recruits three signaling proteins that promote polarized cell growth: Cdc42, Cdc24, and Bem1. Cdc42 is the GTPase described above as an activator of Ste20. Cdc24 activates Cdc42 by inducing the exchange of GDP for GTP. By bringing together Cdc42 with the protein that activates it, Far1 promotes the activation of Cdc42 in the region of the cell membrane containing free $G_{\beta\gamma}$ complexes.

Next, Cdc42, bound to GTP, and Bem1 recruit proteins that promote membrane growth. One such protein is called Bni1. It is a member of a family of proteins called formins, which were first identified because mice lacking such proteins exhibited limb deformities. Studies of Bni1 and other formins have shown that they directly nucleate the polymerization of the cytoskeletal protein actin (Fig. 5-7). In mating yeast cells, Bni1 promotes the formation of actin filaments, which act as tracks for secretory vesicles. The membrane-bound vesicles fuse with the cell membrane to deposit membrane and enzymes at the shmoo tip, causing it to grow. Thus, a series of protein–protein interactions, beginning with the active $G_{\beta\gamma}$ complex, result in the growth of the cell membrane toward the highest concentration of mating pheromone.

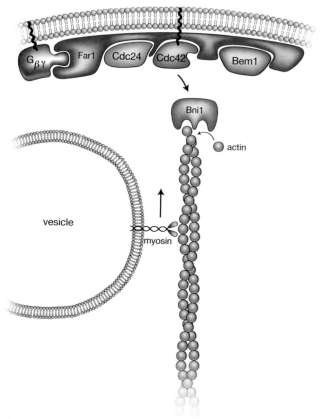

Figure 5-7. $G_{\beta\gamma}$ induces polarized cell growth by promoting actin polymerization. Far1 binds to the free $G_{\beta\gamma}$ complex as well as Cdc24, Cdc42, and Bem1. Cdc24 activates Cdc42 by promoting the exchange of GTP for GDP. Active Cdc42 binds to Bni1, inducing actin polymerization in the vicinity, which leads to the delivery of secretory vesicles. Movement of these vesicles to the plasma membrane requires a "myosin"—this is a so-called "motor protein" that uses the energy of ATP hydrolysis to walk long actin filaments. Not shown are two positive-feedback mechanisms that reinforce cell polarization at the site of highest $G_{\beta\gamma}$ concentration: (1) secretory vesicles deliver additional Cdc42 protein to the plasma membrane, increasing its concentration; and (2) GTP-bound Cdc42 binds Bem1 increasing the affinity of Bem1 for the complex; Bem1 in turn binds to Cdc24, thereby increasing the association of Cdc24 with the complex, which in turn promotes the activation of Cdc42.

Cells Adapt to Mating Pheromone through Negative-Feedback Mechanisms

After prolonged exposure to mating pheromone, **a** and α cells begin to lose their sensitivity to pheromone and will reenter the cell cycle. Such an "adaptation" (or "negative-feedback") is important in the event two cells of the opposite mating type are attracted to one another, but cannot get close enough to mate. Presumably it

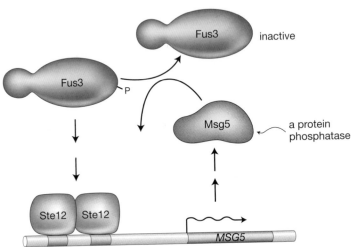

Figure 5-8. Negative feedback of pheromone signaling by induction of a phosphatase that dephosphorylates Fus3.

would be better to give up the attempt rather than to arrest the cell cycle indefinitely in the hope of mating. Several mechanisms are important for adaptation. I will discuss two mechanisms that are well understood. Two genes whose transcription is induced by mating pheromone are *MSG5* and *SST2*.

MSG5 encodes a protein phosphatase that is specific for removing the activating phosphorylation on the Fus3 MAP kinase. Thus, induction of *MSG5* mRNA in response to pheromone signaling, followed by its translation into protein, results in the dephosphorylation of active Fus3 and decreased signaling (Fig. 5-8).

Sst2 inhibits signaling at the top of the hierarchy (Fig. 5-9). It binds to the G_α protein and accelerates the hydrolysis of GTP into GDP, which allows the G_α protein to bind and inhibit the $G_{\beta\gamma}$ complex. In addition to reducing the cells' sensitivity to mating pheromone over time, Sst2 plays another role. This is to prevent cells from erroneously perceiving that mating pheromone is present when it is not: In cells that lack *SST2*, a fraction of the cells arrest their cell cycle and shmoo spontaneously, even in the absence of pheromone.

As with virtually all aspects of the pheromone response pathway, the proteins that mediate adaptation are conserved across many species. Phosphatases homologous to Msg5 are found in humans. Sst2 is the founding member of a family of proteins called RGS proteins (for *r*egulator of *G* protein *s*ignaling) that regulate heterotrimeric G proteins in a variety of organisms. For example, the human protein RGS9 controls the responsiveness of the G_α protein Transducin to light in the retina of the eye. Mice lacking the *RGS9* gene can respond to single flash of light, but not to repeated flashes of light. This is presumably because the inactivation of Transducin after a flash of light occurs too slowly for the mice to see the next flash

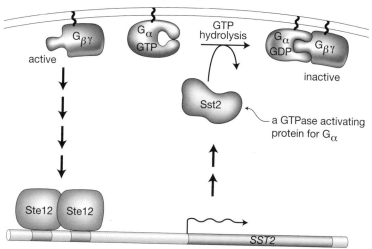

Figure 5-9. Negative feedback of pheromone signaling by induction of a protein that stimulates hydrolysis of GTP that bound to G_α.

of light. These observations provide a vivid example of why it is important for signaling proteins to be able to switch rapidly between their "on" and "off" states.

The Pheromone Response Pathway and the Filamentous Growth Signaling Pathway Share Components, yet Program Distinct Gene Expression Programs

It might seem that signaling components should be dedicated to one pathway. However, this is not the case; rather, components of one pathway are often used by another pathway to respond to a different signal. We have already seen one example of this: Except for the pheromone receptors, **a** and α cells use identical components to respond to the two different pheromones. The responses to the two pheromones are identical except that in **a** cells, transcription of **a**-sgs is increased, whereas in α cells, transcription of α-sgs is increased.

Another example of the sharing of signaling components by pathways that sense different signals is provided by a pathway that mediates the switch to filamentous growth discussed in Chapter 2. Three kinases (Ste20, Ste11, and Ste7), the Ste12 inhibitors, and Ste12 that are involved in the pheromone signaling response are also involved in the induction of filamentous growth, but this pathway activates the transcription of a different set of genes than does the pheromone signaling pathway. Adding synthetic pheromone to cells does not induce these filamentous growth genes, so something must prevent unwanted "cross talk" between the pathways.

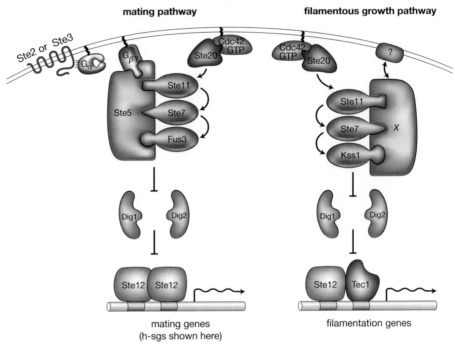

Figure 5-10. The pheromone response pathway and a pathway that controls filamentous growth share multiple components. Not indicated are the following: Dig2 appears to play only a minor role in inhibiting filamentation genes. At many filamentation genes, Ste12 does not contact the DNA directly and Tec1 likely binds cooperatively with another protein.

Figure 5-10 illustrates the pheromone response pathway and the MAP kinase pathway that controls filamentous growth, as we currently understand them. Not all of the components of the filamentous growth pathway are known. For example, it is not known whether there is a scaffold protein analogous to Ste5. What is clear is that this pathway does not involve the pheromone receptor, the G protein, or Ste5.

Two proteins specific to the filamentous growth pathway are known (Fig. 5-10). The first is a MAP kinase called Kss1. Fus3 is not required for filamentous growth, and Kss1 is not required for mating. Thus, the two pathways use different MAP kinases, yet share the three upstream kinases. The second known component specific to the filamentous growth pathway is a transcriptional activator called Tec1 (Fig. 5-10). Tec1 binds to DNA with Ste12 to the promoters of filamentous growth genes. So, in filamentous growth, Ste12 is directed to the appropriate promoters by its association with Tec1. As in the pheromone response, Dig1 and Dig2 act between Kss1 and the Ste12-Tec1 complex.

One might imagine that during mating a shared kinase such as Ste7 could dissociate from the Ste5 scaffold after being activated, gain access to Kss1, and phos-

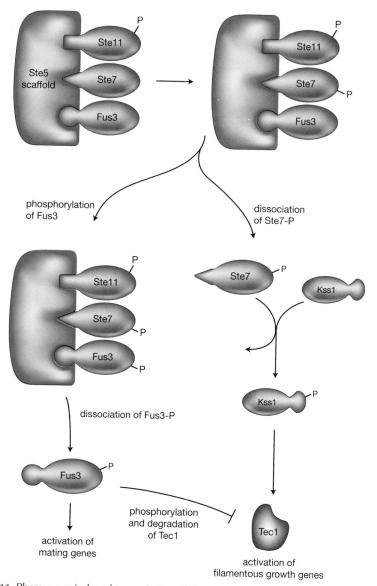

Figure 5-11. Pheromone-induced transcription of filamentous growth genes is prevented by Fus3-induced phosphorylation and degradation of the transcription factor Tec1.

phorylate it (Fig. 5-11). Indeed, this appears to occur because some Kss1 becomes phosphorylated in cells exposed to mating pheromone. In principle, this activated Kss1 should phosphorylate Dig1 and Dig2 and release Ste12 and Tec1 to activate filamentous growth genes, yet in practice these genes are not activated during mating.

The answer to this conundrum turns out to be that activated Fus3 binds and phosphorylates Tec1, the transcription factor specific to the filamentous growth pathway (Fig. 5-11). This phosphorylation targets Tec1 for degradation. So, although some Kss1 is activated during the mating response, the bulk of Tec1 is degraded before its activation by Kss1 can induce filamentous growth genes (Fig. 5-11). Importantly, not all of the Tec1 is degraded, preserving the ability of the filamentous response pathway to respond to its own signals.

Review Articles

Bardwell L. 2005. A walk-through of the pheromone response pathway. *Peptides* **6:** 339–350.

Dohlman H.G. and Thorner J. 2001. Regulation of G protein–initiated signal transduction in yeast: Paradigms and principles. *Annu. Rev. Biochem.* **70:** 703–754.

Schwartz M.A. and Madhani H.D. 2004. Principles of MAP kinase signaling specificity in *Saccharomyces cerevisiae. Annu. Rev. Genet.* **38:** 725–748.

CHAPTER SIX

HOW DIVISION OF A SINGLE CELL CAN PRODUCE TWO CELL TYPES

Asymmetric Cell Division

In Chapter 1, we learned of several principles about how different cell types arise. One of these concerned asymmetric cell division, the process by which a cell divides to produce two cells that differ in their identities and behavior. One general scenario by which an asymmetric cell division occurs is as follows: Copies of a regulator molecule are asymmetrically partitioned prior to cell division such that one of the progeny cells inherits these molecules but the other does not. This inheritance determines whether or not transcription of one or more genes important for the specification of cell type occurs. In this chapter, we will discuss in detail how such an event leads to a change in the genes expressed from the mating-type locus.

Some Strains of Yeast Can Switch Mating Type

In Chapter 3, we learned that mating type is determined solely by which allele is present at the mating-type locus MAT: $MAT\mathbf{a}$ versus $MAT\alpha$. It may therefore come as a surprise to learn that many strains of *Saccharomyces cerevisiae* can switch mating type as they divide. A typical scenario is shown in Figure 6-1. Here we see a mother of mating type α (this cell is a called a mother cell because it gave rise to a daughter in the previous generation). This α mother cell then divides by budding, yielding a mother cell and a daughter cell (defined as a cell that was born as a bud). But the offspring are not α cells; they are $\mathbf{a}$ cells. When the mother cell and daughter undergo a new round of cell division, the daughter cell produces two $\mathbf{a}$ cells, but the mother cell switches again and produces an α mother and an α daughter (Fig. 6-1)!

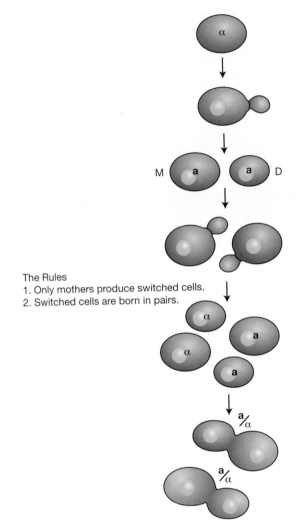

The Rules
1. Only mothers produce switched cells.
2. Switched cells are born in pairs.

Figure 6-1. Mating-type switching.

How does this happen? It turns out that the yeast genome contains one extra copy each of the DNA sequences that encode *MAT*a and *MAT*α. These copies are not transcribed and are known as "silent mating-type cassettes." During the process of mating-type switching, a copy of the silent cassette is produced that encodes the mating type opposite of the one present at the *MAT* locus, and this copy is used to replace the DNA sequence present at *MAT*. As a consequence, the mating type of the cell is changed.

It is thought that mating-type switching exists because, in the wild, diploid cells have a competitive advantage, probably because they divide more rapidly than haploid cells. The example in Figure 6-1 shows how switching helps to produce diploids: After two divisions, there are four cells, two **a** and two α; each of the **a** cells can mate with an α (Fig. 6-1).

The Two Rules of Switching: Only Mother Cells Produce Switched Cells, and Switched Cells Are Born in Pairs

Mating type switching is an orderly process that is governed by two rules.

Rule 1: Only mother cells can produce switched progeny. (Note that in the next generation the original daughter gives rise to a bud, making it a mother cell now. This new mother can now produce progeny of a mating type opposite of its own.)

Rule 2: Switched cells are always born in pairs: That is, a mother cell divides to produce two cells that are both of the mating type opposite of the mother's.

These two rules are illustrated in Figure 6-2, which diagrams the relationship between cell lineage and mating-type switching over four generations. The mating type of the mother cell after cell division is shown on the left of each division and that of the daughter cell is shown on the right. The first rule (only mothers switch) suggests

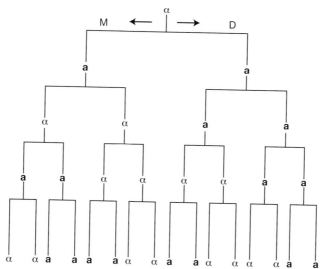

Figure 6-2. Pedigree of mating-type switching. Four generations of cell division are depicted from top to bottom. The horizontal lines indicate cell division. "**a**" and "α" indicate the mating type of the cell produced by a given cell division. The mating type of the mother cell (M) produced by a division is always shown on the left branch, whereas the mating type of a daughter cell (D) produced by a division is shown on the right branch. For example, the original cell at the top of the pedigree was of mating type α. It divided to produce a mother cell and daughter cell, both of which were of mating type **a**.

that the switching machinery is functional only in mother cells. The second rule suggests that mating type is switched before the *MAT* locus has been replicated. We will explore the molecular mechanisms that underlie these two rules later in the chapter.

One can see from the pedigree of mating-type switching shown in Figure 6-2 that the process involves asymmetric cell divisions. That is, each cell division produces two kinds of cells: one capable of producing switched cells (the mother cell) and one incapable of producing switched progeny (the daughter cell).

BOX 6-1. STEM CELLS RENEW TISSUES AND THEMSELVES

The topic of this chapter is asymmetric cell division, and we have seen how yeast can divide asymmetrically to produce a cell that can switch mating types (a mother cell) and a cell that cannot (a daughter cell).

Stem cells can divide in a similar pattern (Fig. 6-3): They can divide asymmetrically, producing a stem cell and a progenitor cell, or they can divide symmetrically, producing two stem cells. In the case of asymmetric division, the progenitor cell either under-

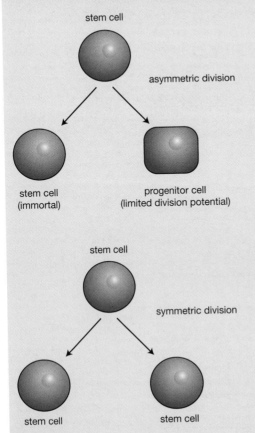

Figure 6-3. Two types of cell division performed by stem cells.

(Continued)

Box 6-1. *(Continued)*

goes immediate differentiation to a particular cell type or is committed to generating a particular cell type, or set of cell types, within a limited number of subsequent divisions.

One of the best-characterized stem cells is the hematopoietic stem cell that occurs in human bone marrow. This type of stem cell gives rise to all human blood cells (red blood cells, lymphocytes, macrophages, etc.) and is used to treat certain cancers of the blood. When chemo- and/or radiotherapy is used to kill rapidly dividing cancer cells, the treatment also kills the patient's stem cells. For the patient to survive, the stem cells must be replaced with cells transplanted from a compatible family member.

More than 25 years ago, research into very early stage mouse embryos showed that primitive stem cells can also be isolated. These cells have the property of being able to differentiate (under the right conditions) into a dazzling variety of cell types in a culture dish. The genomes of the cells can be manipulated in cell culture and the cells can be used to regenerate whole animals (e.g., mice carrying specific engineered mutations). Human embryonic stem cells have also been isolated and these have the potential to be used in the treatment of common but incurable illnesses such as Parkinson's disease. Embryonic stem cells could serve as a source of replacement tissues for cells that are irreversibly lost through the course of such degenerative illnesses.

Box 6-2. In the Fruit Fly, Asymmetric Cell Divisions Produce the Cells of the Sensory Organs

Asymmetric cell divisions are essential for the development of organs. The bristles found on the body of the fruit fly *Drosophila* provide a simple example (Fig. 6-4). These bris-

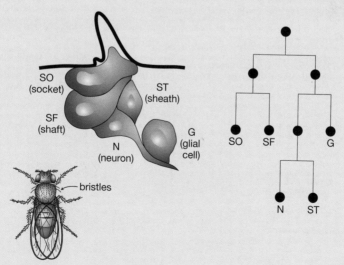

Figure 6-4. Asymmetric cell division produces the bristle organ in *Drosophila melanogaster.* Shown is a drawing of the cell types of a bristle sensory organ. Also shown is how the cell types are produced by a series of asymmetric cell divisions.

(Continued)

tles are actually sensory hairs (and are one of the many reasons it is difficult to swat a fly!). They are composed of five cells that arise from a single cell through four asymmetric cell divisions (Fig. 6-4). One cell makes the shaft of the bristle, another makes the socket that surrounds the shaft, and a third cell forms a sheath around the socket and the shaft. The fourth cell is a neuron that sends sensory information from the hair to the central nervous system of the fly, and the fifth cell is a glial cell, a support cell for the neuron. Thus, just a few asymmetric divisions can produce a sophisticated organ containing five different cell types. Like mating-type switching, these differences are caused by asymmetric inheritance of specific factors during cell division that determine the fate of the cell that inherits them.

Switching Results from Nonreciprocal Recombination between the Mating-type Locus and Silent Mating-type Loci

As discussed above, the ability of yeast to switch mating type repeatedly is dependent on the existence of the silent mating-type cassettes. Like the *MAT* locus itself, these silent loci, called *HMLα* and *HMRa*, are on chromosome III (Fig. 6-5). The two *HM* loci (*HM* stands for "homothallic," meaning "the ability to switch") are silenced by sequences that flank them called E and I, which will be discussed in the next chapter.

During mating-type switching, if *MAT* contains the **a** allele, it is replaced with the α allele from *HMLα* (Fig. 6-6). Likewise, if *MAT* contains the α allele, it is replaced by the **a** allele from *HMRa*. In the process, *HMLα* and *HMRa* are not altered: They merely serve as a template for the conversion of *MAT* to the opposite allele. This explains why switching can occur again and again: The information at *MAT* can be replaced an infinite number of times by the information present at the *HM* loci.

In the remainder of the chapter, we will consider the following three questions. (1) How is the information transferred from the *HM* loci to the *MAT* locus? (2) How

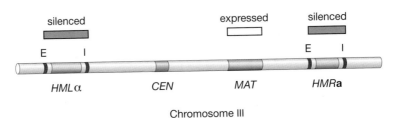

Chromosome III

Figure 6-5. The silent mating-type cassettes *HMLα* and *HMRa* are present on opposite ends of chromosome III.

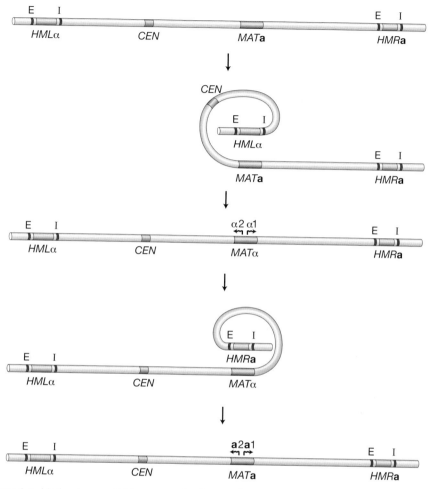

Figure 6-6. Mating-type switching occurs by the transfer of sequence information from the silent cassettes to the *MAT* locus.

is the process regulated to account for the "mother only" and the "switching in pairs" rules? (3) How does the mother cell "know" which *HM* locus contains the information for the opposite mating type?

To understand how switching happens, we must first consider the details of the *MAT* and *HM* loci (Fig. 6-7A). All three loci are made up of three adjacent DNA sequences, called X, Y, and Z. X and Z are identical in all three loci, but Y comes in two versions: Y**a** and Yα. *HMR***a** therefore has the sequence X-Y**a**-Z and *HML*α has the sequence X-Yα-Z. As we will see shortly, the presence of the X and Z sequences at both the *MAT* locus and the silent cassettes is necessary for switching to occur.

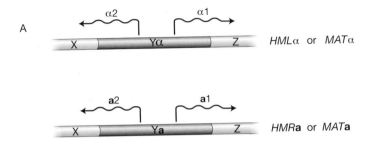

A

α2   α1

HMLα or MATα

a2   a1

HMRa or MATa

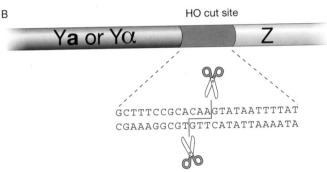

B   HO cut site

Ya or Yα   Z

GCTTTCCGCACAAGTATAATTTTAT
CGAAAGGCGTGTTCATATTAAAATA

Figure 6-7. Sequence elements of **a** and α cassettes. (*A*) Both cassettes contain the common sequences X and Z but differ in the intervening sequence Y. (*B*) The cut site for HO spans the Y–Z junction.

How does the switch work? Mating-type switching occurs by a mechanism similar to one used for repair of damaged DNA. If a cell contains two copies of a DNA sequence, then damage to one of the copies can be repaired by a process called "homologous recombination" in which the cell effectively cuts out the damaged segment and replaces it using information from the undamaged copy. Repair by homologous recombination occurs through the action of many enzymes that have the ability to search the entire genome for sequences identical to one that is damaged, generate a copy of the intact sequence, and use that copy to replace a damaged sequence (the molecular details of this process will be covered below). This type of DNA repair process occurs in diploid cells, which have two copies of each DNA sequence, as well as in haploid cells at a specific point in their cell cycle—namely, when they have undergone DNA replication but not yet divided such that they also have two copies of every sequence.

In the case of mating-type switching, the inciting damage is a double-stranded break that is induced specifically at the *MAT* locus (Fig. 6-7B). The break is repaired by the homologous recombination enzymes mentioned above, which find one of the *HM* loci by searching the genome for sequences identical to the X and Z sequences at *MAT* and then use that silent cassette to repair the damage at the *MAT* locus. In the process of this repair process, the Y region at *MAT* is removed and replaced with a fresh copy of the Y region of the *HM* locus.

A Haploid-specific Endonuclease Initiates Recombination by Producing a Double-stranded Break at the Active Mating-type Locus

What causes the double-stranded break at the *MAT* locus? Figure 6-7B shows a magnified view of this region and highlights a sequence found at the break site. This site is recognized by an endonuclease encoded elsewhere in the genome by the haploid-specific *HO* gene (*HO*, like *HM*, stands for *ho*mothallism). HO is similar to restriction endonucleases, except that its recognition sequence is considerably longer: It recognizes a 24-bp sequence that spans the Y–Z junction and that exists nowhere else in the genome. Although this sequence occurs at both the *MAT* and the *HM* loci (all have the Y–Z junction), only the sequence at *MAT* is cleaved by HO. This is because the factors that prevent transcription of the *HM* loci also prevent cleavage by HO. Because *HO* is a haploid-specific gene, only haploid cells can undergo mating-type switching.

As mentioned above, only some yeast strains can undergo mating-type switching. The reason that some yeast strains (including those commonly used in the laboratory) cannot switch is that they carry mutations in the *HO* gene that inactive it.

Although the exact details of the recombination events that occur after HO cleavage remain unknown, Box 6-3 describes a simplified version of a currently favored model. The basic idea is that unwinding the two strands of an *HM* locus exposes single-stranded DNA that is used as a template for DNA synthesis. This newly synthesized copy of the *HM* sequences is used to repair the double-stranded break at the *MAT* locus. The net result is the replacement of *MAT* with DNA sequences coding for the opposite mating-type locus.

Box 6-3. A Model for Recombination during Mating-type Switching

Figure 6-8 describes the process by which mating-type switching is thought to occur. In the diagram, the top portion of each step shows the silent locus (*HMLα*, in this case) and the lower portion shows the *MAT* locus (*MATa*, in this case). Although the process is diagrammed as occurring in seven sequential steps, several of these likely occur simultaneously in the cell.

Step 1.   Cleavage by HO at *MATa*, degradation of the 5′ strand of Y and Z by exonuclease activity. Exonuclease activity on one side of the break exposes a single-stranded region of the Z sequence, whereas on the other side of the break, the same activity degrades the 5′ strand of Y**a** sequence and continues into the X sequences.

(Continued)

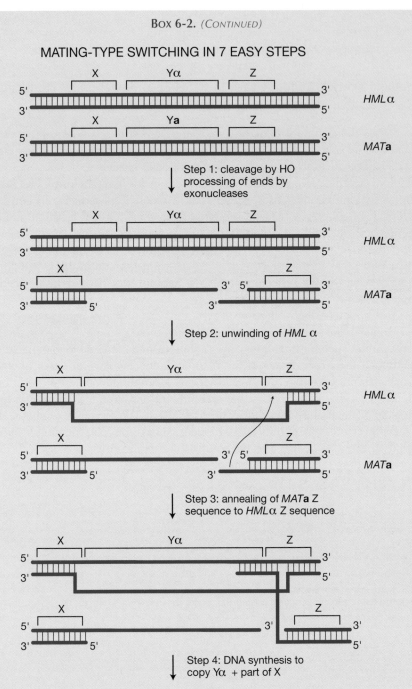

Box 6-2. *(Continued)*

MATING-TYPE SWITCHING IN 7 EASY STEPS

Figure 6-8. Detail of the steps necessary for mating-type switching.

BOX 6-2. *(CONTINUED)*

MATING-TYPE SWITCHING IN 7 EASY STEPS (continued)

Step 4: DNA synthesis to copy Yα + part of X

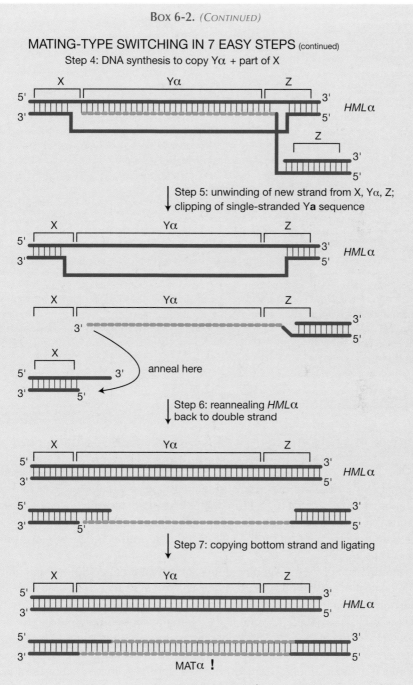

Figure 6-8. *(Continued)*

Box 6-2. (CONTINUED)

Step 2. The two strands of *HMLα* DNA are unwound, exposing part of the X sequence, all of Yα, and part of Z.

Step 3. The single-stranded Z region from *MATa* pairs up with the single-stranded Z sequence in *HMLα*.

Step 4. DNA synthesis proceeds from the 3´ end of the Z sequence of *MATa*, using the Z sequence of *HMLα* as a template. Synthesis continues through the entire Yα region and into part of the X sequence.

Step 5. The newly synthesized DNA and the portion of the *MATa* Z sequence that was base-paired to the Z sequence from *HMLα* are separated ("unwound") from the *HMLα* sequences.

Step 6. The residual single-stranded Ya sequence at *MATa* is clipped off. Then, the 3´ end of the newly synthesized DNA base-pairs with the remaining complementary single-stranded X sequence of *MATa*.

Step 7. Second-strand DNA is synthesized from the 3´ end of the X sequence in *MATa*, using the newly synthesized DNA (generated in Step 4) as a template. Gaps and nicks are repaired using DNA polymerase and ligase activities.

The end result is that the *MAT* locus has been transformed—in this case, from *MATa* into *MATα*. First, *HMLα* was used as a template to synthesize one strand of Yα DNA. That strand was then used as a template to synthesize a second strand. The primers for these synthesis steps came from the resected Z sequence in *MATa* and the resected X sequence from *MATa*, respectively. This process is known as "synthesis-dependent strand annealing."

Switched Cells Are Born in Pairs Because the *HO* Is Transcribed Only in the G₁ Phase of the Cell Cycle

Now that we understand the events of mating-type switching, we can look at its regulation. It turns out that regulation of the time and place of *HO* gene expression explains the regulation of the whole process. Specifically, the *HO* gene is transcribed only transiently during the cell cycle—late in the G_1 phase—and only in mother cells.

Transcription of the *HO* gene requires an activator called SBF, which binds to multiple sites in a region of the promoter called *URS2* (Fig. 6-9). Because SBF is active only briefly at the end of the G_1 phase of the cell cycle (for reasons described in detail in Fig. 6-9), the *HO* gene is therefore transcribed only transiently before returning to an inactive state. This brief pulse of transcription suffices, however, for just enough of the HO protein (which itself has a short half-life) to be synthesized to produce a single double-stranded break at the *MAT* locus before the locus is replicated during S phase.

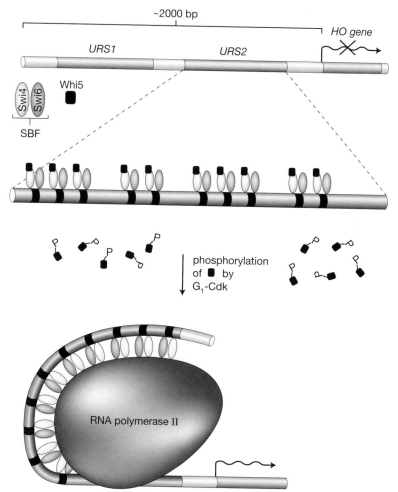

Figure 6-9. A cell cycle–regulated transcription factor called SBF induces transcription of *HO* late in G₁ phase. SBF is inhibited by a protein called Whi5. Whi5 is inactivated by phosphorylation by protein kinases that control the G₁ phase of the cell cycle and that are only active late in G₁. These are indicated by "G₁-Cdk" in the figure. Once Whi5 is phosphorylated, SBF recruits RNA polymerase II and associated factors, allowing transcription to occur.

Switching Occurs Only in Mother Cells Because a Repressor of the *HO* Gene, Ash1, Is Present Only in the Nuclei of Daughter Cells

The cell cycle regulation of the *HO* promoter is superimposed on a second mechanism: The "mother only" expression of HO that requires a second region of the *HO* promoter called *URS1*. At the end of this section, and in the next section, we

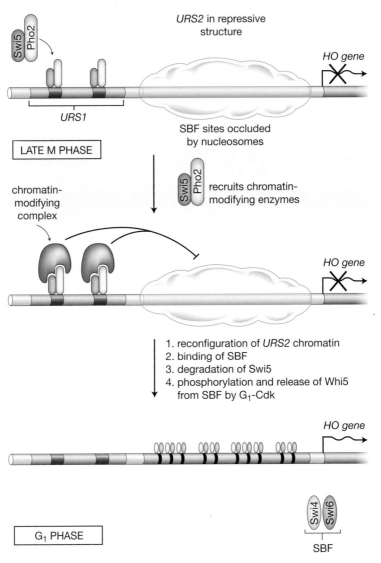

Figure 6-10. Binding of SBF to *URS2* requires that Swi5-Pho2 induces a change in repressive chromatin (shown as a cloud around *URS2*) so that SBF can access its sites. For simplicity, RNA polymerase II is not shown.

will arrive at how the "mother only" pattern occurs, but first it is necessary to discuss Swi5 and Pho2, two proteins that bind to *URS1* (Fig. 6-10).

Swi5 binds to sites in *URS1* cooperatively with Pho2. Swi5 is tightly regulated by the cell cycle; its gene is first transcribed late in G_2 phase and the protein enters

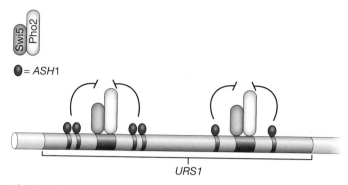

Figure 6-11. The daughter-specific repressor of *HO*, Ash1, suppresses the action of Swi5-Pho2.

the nucleus as nuclear division is occurring. Swi5 binds to the *HO* promoter for only a short period of time (5–10 min). However, it has been shown that this brief binding must occur for the *HO* promoter to be activated subsequently by SBF. This is because Swi5-Pho2 recruits enzymes that modify the structure of the promoter (in ways that are not understood in detail) that renders the target sequences in *URS2* accessible for binding by SBF (Fig. 6-10). These enzymes affect the conformation of general DNA-packaging proteins called histones in the region of *URS2* (histones and chromatin—the ensemble of DNA and its associated proteins—will be discussed in detail in Chapter 7). Note that once bound to DNA, SBF remains inactive until late in G_1 phase (Fig. 6-9).

The "mother only" pattern occurs because there is a repressor of HO transcription, called Ash1. Ash1 accumulates after nuclear division to a much greater extent in the daughter nucleus than in the mother cell nucleus. Because of high levels of Ash1 in the daughter cell nucleus, expression of the *HO* gene is repressed in the daughter cell in the next G_1 phase. Ash1 blocks *HO* expression by interfering with the chromatin-modifying activities recruited by Swi5-Pho2 (Fig. 6-11).

AshI Is Transcribed in Both Mother and Daughters, but Its Messenger RNA Is Transported to Daughters along Actin Filaments by a Myosin Motor

The Ash1 protein accumulates specifically in the daughter cell nucleus prior to cell division. This occurs because, although the *ASH1* gene is transcribed in both mother and daughter nuclei, all of the *ASH1* messenger RNA (mRNA) is transported to the tip of the daughter cell bud (Fig. 6-12). The sequence of events is as follows.

The *ASH1* gene itself is transcribed in a cell cycle–regulated fashion. Swi5 also activates *ASH1* transcription just as nuclear division is occurring (Fig. 6-12). Unlike the case at the *HO* promoter, the binding of Swi5 to the *ASH1* promoter immedi-

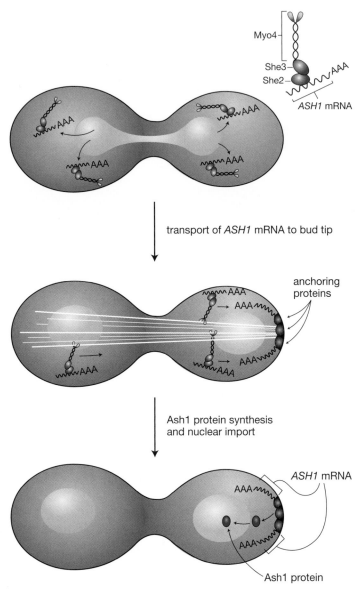

Figure 6-12. Ash1 accumulates selectively in the daughter nucleus prior to cell division because its mRNA is transported to the tip of the daughter bud.

ately induces the transcription of the *ASH1* gene. The resulting *ASH1* transcript is handled differently from most cellular mRNAs. Before the mRNA is exported from the nucleus, specific signals in the Ash1 mRNA are recognized by a protein called She2 (Fig. 6-12). In the cytoplasm of the cell, two other proteins—She3 and

Myo4—then associate with the She2-*ASH1* mRNA complex. Myo4 is a myosin, a type of motor protein that binds to and walks along actin filaments. It turns out that the growth of the bud (like that of the shmoo) is achieved through the transport of vesicles along oriented actin filaments that lead to the tip of the bud. Through Myo4, the *ASH1* mRNA binds to these filaments and is transported to the bud tip (Fig. 6-12). When it arrives, the mRNA is anchored to the tip of the bud by several proteins. This is how the *ASH1* mRNA, which is synthesized by transcription in both mother and daughter nuclei, is asymmetrically localized prior to cell division. The Ash1 protein is then translated and accumulates selectively in the nearby daughter cell nucleus, preventing subsequent activation of the *HO* gene in the daughter cell.

Taking a step back, we can now understand why the events that determine whether the *HO* gene is transcribed during the late part of G_1 are set up in the previous M phase: The repression of the *HO* gene in daughter cells requires the asymmetric partitioning of the *ASH1* mRNA into the bud prior to cell division.

Switching Is Directional

As mentioned earlier, during mating-type switching, DNA at the *MAT* locus is nearly always (~90% of the time) replaced by DNA of the opposite allele. This prevents futile cycles of recombination, which would limit the efficiency of switching. As we will learn, the mechanism that controls the direction of switching requires *HMLα* and *HMR***a** to be at opposite ends of chromosome III. The first clue to the mechanism is that the rate of recombination in the entire left arm of chromosome III is extremely low. It is hypothesized that this arm of the chromosome contains sequences that bind recombination-suppressing proteins. Therefore, all other things being equal, a double-stranded break at *MAT* will most likely (~90% of the time) be repaired using sequences at *HMR***a**, and this is indeed what happens in *MATα* cells. But how is it that *MAT***a** is most often replaced by *HMRα* in **a** cells?

Directionality Is Controlled by a Recombination Enhancer That Promotes the Use of the *HMLα* Silent Cassette in *MAT***a** Cells

The answer is that, in *MAT***a** cells, there is a mechanism that stimulates recombination across the left arm of chromosome III to such an extent that recombination between *MAT* and *HMLα* occurs approximately 90% of the time (Fig. 6-13A). This mechanisms involves a so-called "recombination enhancer" (*RE*), a 250-bp sequence that lies on the left arm of chromosome III and greatly increases the efficiency recombination between *MAT* and *HMLα*. The *RE* is active only in **a** cells and is the reason why *HMLα* is chosen over *HMR***a** as the donor of information when **a** cells switch mating type.

RE stimulates recombination between *MAT* and any homologous sequences placed anywhere on the left arm of the chromosome, indicating that it influences

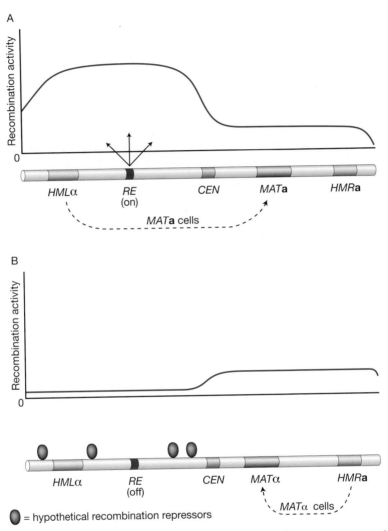

Figure 6-13. The directionality of mating-type switching is controlled by a recombination enhancer (*RE*) that is active only in **a** cells.

recombination over some distance. A protein called Fkh1 binds to the *RE* in **a** cells (Fig. 6-14). This protein, which in other situations functions as a transcriptional activator, is required for the high levels of recombination along the left arm of chromosome III seen in **a** cells. However, its mechanism of action is unknown. What is clear from the greatly enhanced rates of recombination with *HMLα* in **a** cells compared to α cells is that Fkh1 does more than merely overcome the hypothesized recombination inhibitors on the left arm of chromosome III.

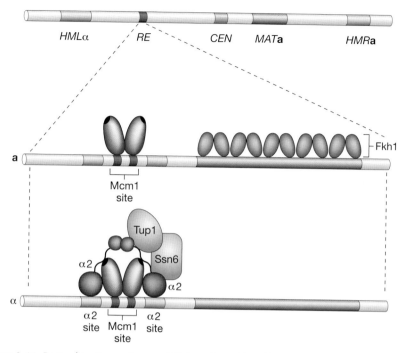

Figure 6-14. Recombination enhancer (*RE*) is activated by Fkh1 and repressed by α2-Mcm1.

In *MAT*α Cells, α2 Binds to and Inactivates the Recombination Enhancer

In *MAT*α cells, the activity of the *RE* is suppressed to ensure that *HMR*a is used as the donor during mating-type switching (Fig. 6-13B). The inactivation is achieved by α2, which binds cooperatively with Mcm1 at the *RE* (Fig. 6-14). Inactivation of the *RE* also requires that α2 recruit the Tup1-Ssn6 corepressor that we encountered in Chapter 3. This suggests that transcription in the *RE* region might regulate its function. Indeed, it has been shown that there is transcription in **a** cells, but not α cells, within the *RE* region.

Review Articles

Cosma M.P. 2004. Daughter-specific repression of *Saccharomyces cerevisiae* HO: Ash1 is the commander. *EMBO Rep.* **10:** 953–957.

Haber J.E. 1998. Mating-type gene switching in *Saccharomyces cerevisiae*. *Annu. Rev. Genet.* **32:** 561–599.

Wittenberg C. and Reed S.I. 2005. Cell cycle–dependent transcription in yeast: Promoters, transcription factors, and transcriptomes. *Oncogene* **24:** 2746–2755.

CHAPTER SEVEN

GENE SILENCING IN THE CONTROL OF CELL TYPE

In the last chapter, we were introduced to the silent cassettes, *HMR***a** and *HML*α, which act as donors of DNA sequence information during mating-type switching. In this chapter, we will learn how these DNA regions are kept silent.

Before delving into the details, let us define "silencing" as a mechanism that prevents the accumulation of messenger RNA (mRNA) from a region of DNA (which may contain more than a single gene) and works independently of the usual regulatory sequences of the individual genes (Fig. 7-1). As such, silencing differs from transcriptional repression (such as that mediated by α2), which acts on individual genes through a specific promoter element.

We will learn that silencing is initiated at specific sites and spreads from those sites by changing the structure of chromatin (we will define chromatin here as DNA and its associated proteins). Although silencing can spread laterally over long stretches of DNA, its progression is inhibited by a number of mechanisms that protect active regions of the chromosome from inappropriate silencing. These three key features of silencing in yeast—the involvement of changes in chromatin structure, the ability to spread, and the existence of mechanisms to limit its spread—also characterize silencing in other eukaryotes, including humans.

Silencing Involves Changes in the Modification Patterns of Histone Tails

Understanding the relationship between chromatin structure and silencing requires that we first discuss the composition of chromatin. DNA in eukaryotes is packaged into nucleosomes. A nucleosome consists of a segment of DNA wrapped around a core of eight polypeptide subunits the "octamer." This core is composed of two copies each of four different histone proteins: H2A, H2B, H3, and H4. Protruding from this octamer core are the unstructured amino-terminal tails of the histones

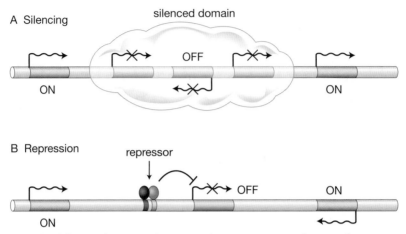

A Silencing

silenced domain

OFF

ON

ON

B Repression

repressor

OFF

ON

ON

ON

Figure 7-1. The difference between silencing and repression. (*A*) A domain of DNA containing three genes that are all shut off by a silencing mechanism. The cloud indicates a change in the chromatin that causes silencing, the details of which are unknown. (*B*) A repressor bound to the regulatory sequences of a gene turn that gene off. Note that repressors can inhibit transcription through changes in chromatin structure and these may or may not be related to those that operate during gene silencing.

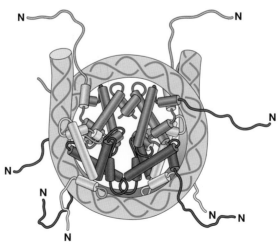

Figure 7-2. The structure of the nucleosome showing the amino-terminal tails protruding from its core. The nucleosome is made up of two copies each of H2A (*light gray*), H2B (*dark blue*), H3 (*light blue*), and H4 (*dark gray*).

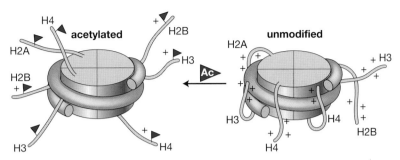

Figure 7-3. Acetylation of specific lysines on the histone tails.

(Fig. 7-2), which can be covalently modified at specific amino acid residues. For example, the tails contain conserved lysine residues that can be covalently modified by the addition of an acetyl group by enzymes called histone acetyltransferases (Fig. 7-3). Histone H4, for example, can be acetylated on lysines at positions 5, 8, 12, and 16 on its amino-terminal tail.

It turns out that the tails of all of the histones at *HMR***a** and *HML*α are deacetylated. Removal of the amino-terminal tail of either H3 or H4 causes a complete loss of silencing, and mutation of specific lysines in the tails that are subjected to acetylation can disrupt silencing. Thus, the tails of histones H3 and H4 and their state of acetylation are important for silencing.

Experiments performed with purified chromatin have shown that acetylation of one of these residues, lysine 16, can cause chromatin to become less compact, probably by destabilizing interactions between nucleosomes. Acetylation of tails can also either promote or block the ability of proteins to bind the nucleosome. It has been suggested that these effects of acetylation are relevant to silencing.

DNA Sequences Called Silencers Flank Each Silent Mating-type Cassette and Promote Silencing

As mentioned in Chapter 6, the *HMR***a** and *HML*α loci are flanked by DNA sequences called "silencers," which are not present at the *MAT* locus (Fig. 7-4).

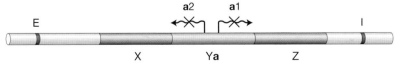

Figure 7-4. The *HMR***a** silent mating-type cassettes showing the E and I silencers.

These are regulatory sites similar to ones we encountered in Chapter 3; as we will see, their function is to recruit proteins necessary for silencing. We will focus here on the silencers that flank *HMR***a**; the silencers that flank *HML*α work in a similar fashion. To the left of the XY**a**Z region of *HMR***a** is the E silencer (so named because it is "essential" for silencing). To the right is the I silencer ("*i*mportant" for silencing). As their names would suggest, deleting the E silencer results in a complete loss of silencing, whereas deleting the I silencer results in a partial loss of silencing.

The silencers are constructed from shorter elements (Fig. 7-5). These elements are important for the function of the silencers, but appear to act redundantly with

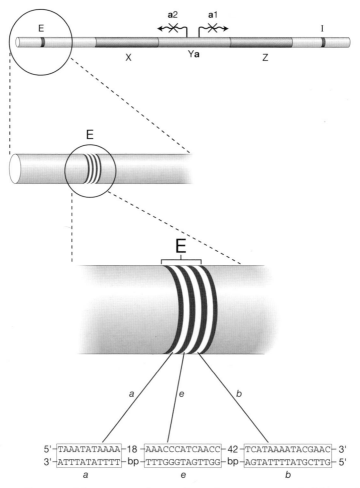

Figure 7-5. Close-up of the *HMR***a** E silencer shows that it is composed of three subelements.

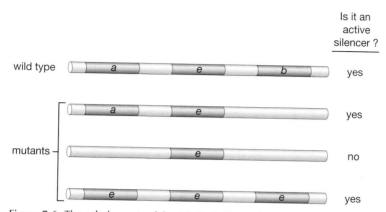

Figure 7-6. The subelements of the *HMR***a** E silencer have equivalent functions.

each other. Any one of the three sites at the *HMR***a** E silencer can be deleted with no loss of silencing, but deletion of any two results in the elimination of silencing. Furthermore, the sites in the *HMR***a** E silencer can be replaced with three copies of just one of the elements, for example, and still function as a silencer (Fig. 7-6). Thus, the silencers are made up of a number of subelements that each have a similar function, and at least two subelements are required for a silencer to function.

These three elements are recognized by three different proteins (Fig. 7-7). Each of these proteins also has one or more other functions in the cell. One site is recognized by the origin recognition complex (ORC), a protein complex that binds to sites, called origins, where DNA replication is initiated on the chromosome. The other two sites are bound proteins called Rap1 and Abf1. These function as activators of transcription when bound to single sites in promoters. Rap1 also has a third function at telomeres (more on that toward the end of the chapter). Current evidence suggests that each protein evolved its other role(s) before acquiring its ability to promote silencing.

Figure 7-7. The subelements of the *HMR***a** E silencer each bind to a different protein.

BOX 7-1. SILENCING AND CANCER: ACUTE LEUKEMIA IS CAUSED BY
A DEFECT IN SILENCING OF *HOX* GENES

As discussed in Chapter 3, α2 and **a**1 are members of the homeodomain family of transcription factors that control cell type in many different species, including humans. In humans, the *HOX* genes represent a large family of homeodomain proteins that operate during embryonic development. Rather than being randomly distributed across the genome, these genes occur in clusters along the chromosome. As development proceeds and tissues mature, the expression of *HOX* genes is gradually reduced. In fully differentiated tissues, *HOX* clusters are silenced via a mechanism similar to the silencing of the *HM* silent cassettes in yeast (this is probably explains, at least in part, why they occur in clusters—so that they can be subjected to gene silencing).

The silencing of *HOX* genes turns out to be relevant to an important class of human cancer. In a large fraction of acute leukemias (which are cancers of white blood cells), this silencing is disrupted, resulting in the inappropriate expression of certain *HOX* clusters (Fig. 7-8). It is now well established that this loss of silencing promotes the formation of leukemia by directing cells to take on an inappropriate proliferative cell fate characteristic of immature cells. This finding has been exploited for the treatment of these tumors. For some leukemias, drug treatments that promote the differentiation of the tumor cells (and consequently the resilencing of *HOX* genes) have been shown to be effective.

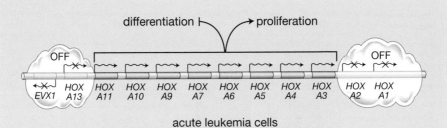

Figure 7-8. Some forms of acute leukemia are caused by loss of silencing of a cluster of *HOX* genes.

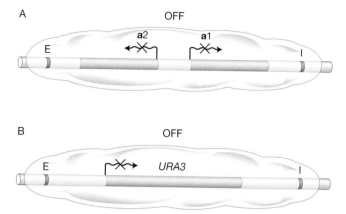

Figure 7-9. Genes other than those encoding **a**1 and **a**2 are silenced when placed within *HMR*. (*A*) The normal situation. (*B*) Placement of a typical gene used to test for silencing (*URA3*) within *HMR***a** results in its silencing.

Many Different Genes Can Be Silenced by Placing Them in Silent Cassettes

As shown in Figure 7-4, the promoter regions of the **a**1 and **a**2 genes lie in the middle of *HMR***a**, away from the E and I silencers. Similarly, the α1 and α2 promoter regions lie in the middle of *HML*α. Therefore, the silencers have to act at a significant distance from the promoters that they affect. This raises the question of whether there is something special about the promoters in *HMR***a** and *HML*α that make them susceptible to silencing. This is not to be the case, as it has been found that placing any one of a number of genes in the region between the E and I silencers results in the gene's transcriptional silencing (Fig. 7-9). Thus, the silencers work over long distances and appear to be able to silence any gene placed between them.

The Sir Complex Mediates Silencing by Spreading from the Silencers into the *HM* Cassettes

When bound to silencer subelements, ORC, Rap1, and Abf1 recruit a three-protein complex called the Sir complex ("Sir" stands for "*s*ilent *i*nformation *r*egulators"). This protein complex spreads along chromatin from the E and I silencers and deacetylates the tails of histones as it goes (more on the exact mechanism in the next section). Two of the proteins in the Sir complex, Sir3 and Sir4, bind directly to the tails of histones H3 and H4.

In addition to associating with the other Sir proteins, Sir4 can also associate with itself, and with Rap1, Abf1, or ORC (i.e., each of the latter three can directly

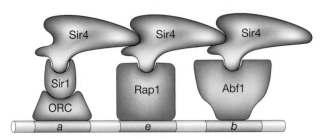

Figure 7-10. Each of the proteins that binds to a subelement of the E silencer binds Sir4, which also interacts with itself.

bind to Sir4). Figure 7-10 shows a model of how weak interactions between each of these three proteins and Sir4 lead to the assembly of the Sir complex at the E silencers. In essence, three copies of Sir4 can bind cooperatively to the E silencer, because it has the ability to bind to proteins bound to each of the subelements of a silencer and an affinity for itself. Such a mechanism is entirely analogous to the cooperative binding of proteins to DNA discussed in Chapter 3, except in this case Sir4 binds proteins that bind DNA, rather than to DNA directly.

A Component of the Sir Complex, the Sir2 Protein, Is a Histone Deacetylase

As described above, it is the interaction between Sir4 and proteins that bind to the subelements of the silencer that initiates the silencing process. However, the spreading of the Sir complex requires the local deacetylation of histone tails. The reason for this requirement is that although Sir3 and Sir4 bind to histone tails, their binding is inhibited by acetylation of the tails. Therefore, for spreading to occur, the Sir complex must first deacetylate nearby tails. The Sir2 component of the complex is a histone deacetylase enzyme capable of removing the acetyl groups. Thus, binding of the Sir complex to the silencers results in local deacetylation of histone tails, which, in turn, allows binding of additional copies of the Sir complex (Fig. 7-11) directly to the deacetylated tails. These complexes then deacetylate further histones, which then bind more of the complex, and so on. Cycles of this process, coupled with protein–protein interactions between adjacent Sir complexes (these have yet to be understood in detail), result in the spread of the Sir complex and conversion of the silent cassettes from an active state containing acetylated histone tails into a repressed state containing deacetylated histone tails and a "coating" of the Sir complex. The resulting silenced domain is refractory to the binding of RNA polymerase II (Fig. 7-12). It is not clear whether it is the state of the histone tails (and consequent effects on either chromatin compaction or the recruitment of proteins), the recruitment of the Sir proteins, or both that ultimately cause silencing.

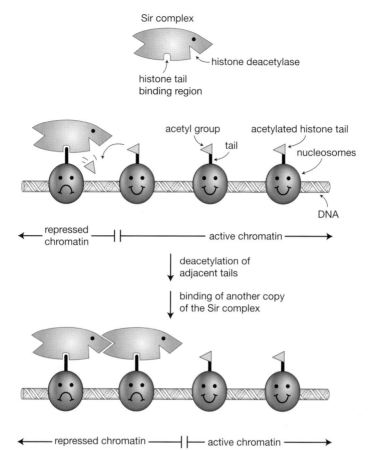

Figure 7-11. The Sir complex spreads along chromatin through cycles of histone tail deacetylation followed by binding to the deacetylated tails.

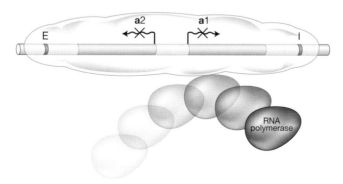

Figure 7-12. Silencing at *HMR*a prevents RNA polymerase II from binding to the **a**1-**a**2 promoter region, but the exact mechanism is unknown.

Several Mechanisms Prevent the Excessive Spread of Silencing

The impression we have up to this point is that the Sir complex spreads from the silencers inward toward the center of the silent cassettes. So what prevents the complex from spreading outward? Such spreading could be disastrous; the silencing of an essential gene would result in the death of the cell. Therefore, cells have evolved several mechanisms to prevent the spread of the Sir complex beyond the borders of the silent cassettes.

One mechanism uses a "boundary element." These are defined simply as DNA sequences that reduce the spreading of silencing. Boundary elements are found on both sides of *HMR***a** and to the left of *HMLα*. Boundary elements are also found in higher organisms, including humans, where again they serve to prevent the spread of silencing across chromosomes.

Boundary elements are not absolute barriers to the spread of silencing, necessitating additional mechanisms. Indeed, there are several chromatin modifications found throughout active chromatin that antagonize silencing (Fig. 7-13). First, another modification of lysine residues on histones, called methylation, can interfere with silencing, depending on which lysine is methylated. In particular, methylation of lysines 4 and 79 on histone H3 is particularly effective in blocking the spread of silencing (these same residues are methylated in the human version of histone H3, suggesting a conserved function). Second, a variant of histone H2A called H2A.Z (also found in human cells) replaces H2A in the promoters of genes in active chromatin; this too inhibits the spread of silencing. The deposition of H2A.Z in promoters is targeted by regulators that bind specific DNA elements (their action is inhibited in silent chromatin), whereas the histone methylations described above occur in the body of the genes in response to transcription induced by activators.

The precise mechanisms by which boundary elements and active chromatin modifications antagonize silencing in yeast is not known.

Silencing near Telomeres Is Similar to *HM* Silencing and Requires the Sir Complex

Silencing in yeast is not limited to the *HM* cassettes. DNA at the ends of chromosomes is also packaged into silent chromatin. In this case, silencing is initiated at the telomeric repeats found at the end of all chromosomes and spreads into the body of the chromosome. As a consequence, this form of silencing becomes weaker with distance from the telomere. In general, tight silencing (i.e., silencing that can block the expression of an otherwise highly transcribed gene) occurs only within a few thousand base pairs of the telomere.

Silencing at telomeres is nucleated by Rap1, the protein mentioned above that helps to initiate silencing at the *HM* cassettes (Fig. 7-14). In the telomeres, Rap1 binds specifically to the telomere repeats and, as at the *HM* silencers, binds to the

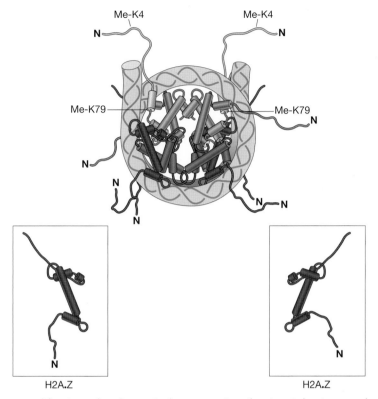

Figure 7-13. Modifications of euchromatin that antagonize silencing. Subunits are colored as in Fig. 7-2 except that H2A.Z rather than H2A is represented by *dark gray*.

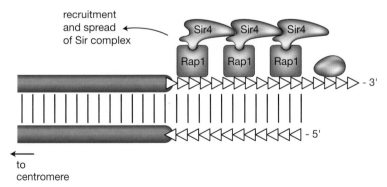

Figure 7-14. Rap1 nucleates silencing at telomeres, which are the ends of chromosomes. Triangles depict short repeated sequences that are found at telomeres. In addition to Rap1, other proteins bind to telomeric repeats.

Sir4 protein, recruiting the Sir complex that then spreads along the chromatin away from the telomere repeats. The binding of Rap1 also has other roles at telomeres, such as helping to control their length and protecting them from degradation.

Telomeric Silencing Is Unstable but Heritable, Producing Heterogeneous Populations of Yeast

At the *HM* cassettes, all cells in a population are silenced. In contrast, silencing at telomeres is less efficient resulting in mixed populations of yeast cells. For a given gene near a telomere, in some cells this gene will be expressed, whereas in others it will be repressed. This bimodal form of silencing has an interesting property: Cells in which a gene near a telomere is repressed tend to give rise to progeny cells in which that gene is also repressed. Likewise, cells in which a gene near a telomere is expressed tend to produce progeny in which that gene is also expressed. Only occasionally will a cell give rise to progeny that have switched the state of gene expression from on to off or off to on.

This switch can be seen in a simple colony assay on an agar plate. This assay takes advantage of a yeast strain in which a gene placed near a telomere causes a color change in the cell, depending on whether that gene is expressed or not (Fig. 7-15). *ADE2* is such a gene. Cells that express *ADE2* are white, whereas those that do not are red (Ade2 is involved as an enzyme required for the biosynthesis of the nucleotide adenine and, when it is missing from cells, an intermediate that produces a red pigment accumulates).

When the *ADE2* gene is placed near a telomere, the resulting colonies are either red with white sectors or white with red sectors. A switch to silencing of the *ADE2* gene produces a red sector in a colony originating from an *ADE2*-expressing cell (Fig. 7-15). Each red sector represents the progeny of a single cell in which the *ADE2* gene was silenced. Occasionally, the repressed *ADE2* gene becomes reactivated, resulting in a white sector within a red sector (Fig. 7-15).

The inheritance of expression states of genes near yeast telomeres is an example of "epigenetic inheritance." This broad term refers to any change in a cell's phenotype that is not caused by a mutation but is nevertheless inherited in the absence of the signal or event that induced the change. The epigenetic inheritance of Sir-dependent silencing is hypothesized to occur through the inheritance of silent chromatin itself (defined as the ensemble of the proteins that recognize the silencers, the Sir proteins, histones, and DNA). While the details of this process are not understood, it is known that after DNA replication, each daughter DNA molecule contains a 50:50 mix of preexisting histones (which appear to be locally placed back onto the DNA after its replication) versus newly synthesized histones. If the "old" histones were to remain deacetylated long enough for Sir complexes to rapidly reassociate with them, it could be that the local deacetylation and spread of the complex described above could then reestablish the silent chromatin state before gene

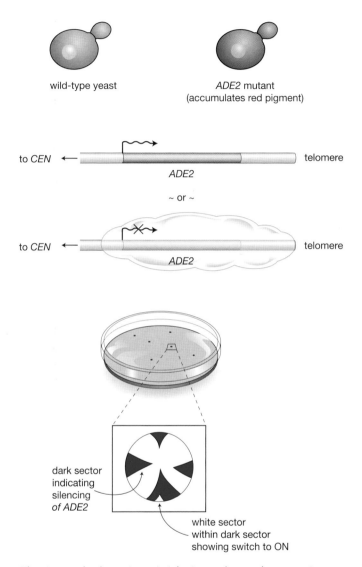

Figure 7-15. Silencing can lead to epigenetic inheritance that can be seen using a sectoring assay.

activation could occur. It is thought that switching from the "off" to the "on" state occurs when, by chance, regulatory proteins "gain the upper hand" to induce the deposition of H2A.Z and histone methylations that, in turn, antagonize silencing. One possibility is that the period when DNA–protein interactions are disrupted during DNA replication is a time that the cell would be vulnerable to such changes.

Yeast cells in a population are a mosaic of cells with different expression states of genes near telomeres. Because there are several hundred genes near telomeres, there is the potential for telomeric silencing to generate substantial diversity. What might be the function of this heterogeneity in the population? In yeast, genes near telomeres tend to be involved in the assimilation of nutrients such as sugars. Recent theoretical work suggests that in environments that are constantly changing, it is advantageous to have a heterogeneous population. This allows an organism to "hedge its bets." For example, by displaying a variety of sugar transporters with different specificities on the surfaces of individual cells, a population can guard against a rapid shift in the type of sugars available, which would otherwise cause the entire population to arrest growth (Fig. 7-16). With heterogeneity in the population, there

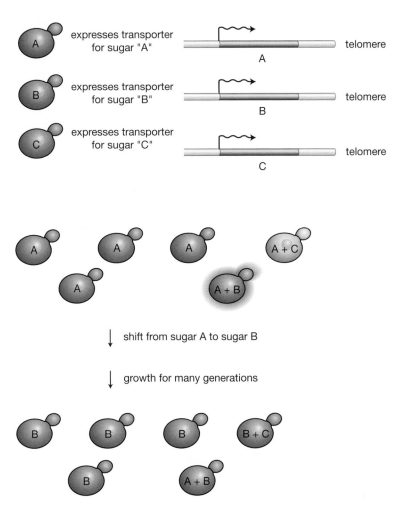

Figure 7-16. Silencing of genes near telomeres can produce heterogeneity in the population that can be advantageous in a changing environment.

will always be some individuals that express the appropriate transporter for the new environment, allowing the population to survive when the environment changes. This strategy is more efficient than requiring, for example, that all the cells in the population express transporters for all potentially available sugars.

Heterogeneity in gene expression resulting from gene silencing near telomeres is not unique to yeast. It occurs in organisms as diverse as the malaria parasite *Plasmodium falciparum* and humans and has serious implications for human disease. In *P. falciparum*, which infects red blood cells, genes encoding surface antigens are clustered near telomeres and subject to silencing by a protein homologous with Sir2 that is encoded by the parasite genome. Variable expression of these surface antigens results in variation in the antigens displayed on the surface of infected red blood cells in people suffering from malaria. This "moving target" poses a difficult challenge for the immune system and a roadblock to the development of an effective vaccine for this deadly disease.

Box 7-2. Silencing of the Silent Mating-type Cassettes in the Fission Yeast *Schizosaccharomyces pombe* Requires Histone Methylation in Addition to Histone Deacetylation

So far we have been discussing *Saccharomyces cerevisiae*, but some other yeasts also undergo mating-type switching that involves silent cassettes. Comparative studies of different yeast types have provided interesting insights into the different ways such systems can be set up. For example, in the well-studied fission yeast *Schizosaccharomyces pombe*, both silent cassettes are in a single 20-kb domain of silent chromatin adjacent to the mating-type locus (Fig. 7-17). Silencing is nucleated by two silencer elements called *cenH* and *REIII* that lie between the two silent cassettes. Although we cannot get into the details here, specific proteins bind to these regions and recruit silencing complexes that then spread in both directions limited only by boundary elements that define the borders of the 20-kb region (Fig. 7-17). Again, histone deacetylation is required for the silencing to spread, but, unlike in *S. cerevisiae*, the spreading process in *S. pombe* is also dependent on a specific histone methylase (Fig. 7-18). In particular, a histone methyltransferase called Clr4 is recruited by factors that bind the *cenH* and *REIII* silencers. Clr4 specifically methylates lysine 9 on the amino-terminal tail of histone H3. This methylation requires lysine 9 to be deacetylated first (a lysine cannot be acetylated

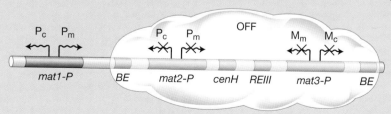

Figure 7-17. Silencing of the silent mating-type cassettes in *Schizosaccharomyces pombe*.

(*Continued*)

BOX 7-2. (*CONTINUED*)

and methylated at the same time). The methylated lysine 9 forms a binding site for a protein called Swi6, which then recruits histone deacetylases (including a homolog of Sir2) and more Clr4. Thus, as compared to *S. cerevisiae*, there is an extra step—methylation—in the spreading cycle of silencing. Because methylation of lysine residues appears to be more difficult to reverse than acetylation, it may serve to enhance the stability of the silenced state. This system of silencing is conserved across evolution. For example, humans have homologs of Clr4 and Swi6 that are involved in silencing. Thus, comparative studies of different yeasts can provide new insights into fundamentally important mechanisms.

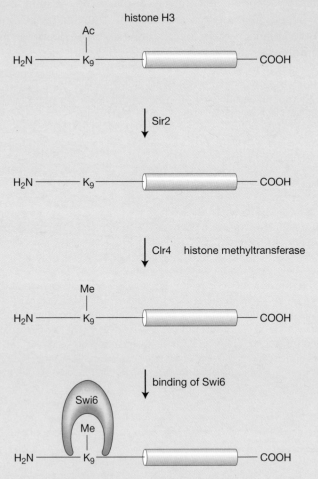

Figure 7-18. Deacetylation and methylation of histone H3 in lysine 9 and its recognition by Swi6 are required for silencing in *Schizosaccharomyces pombe*.

Review Articles

Raisner R.M. and Madhani H.D. 2005. Patterning chromatin: Form and function of histone H2A.Z variant nucleosomes. *Curr. Opin. Genet. Dev.* **16:** 119–124.

Rusche L.N., Kirschmaier A.L., and Rine J. 2003. The establishment, inheritance and function of silenced chromatin in *Saccharomyces cerevisiae*. *Annu. Rev. Biochem.* **72:** 481–516.

van Leeuwen F. and Gottschling D.E. 2002. Genome-wide histone modifications: Gaining specificity by preventing promiscuity. *Curr. Opin. Cell Biol.* **14:** 756–762.

CHAPTER EIGHT

CONTROLLING CELL POLARITY

The proper development and maintenance of multicellular organisms usually requires careful control of the plane of cell division (see Box 8-1 for an example). As we discussed briefly at the end of Chapter 2, yeast cells do not bud at random locations during the cell cycle; rather, the patterns of budding are carefully programmed. Moreover, the patterns differ between the haploid cell types and the **a**/α cell type. Haploid **a** and α cells bud in an axial pattern in which budding occurs adjacent to the site of the previous bud. In contrast, **a**/α cells bud in a more complex bipolar pattern: An **a**/α daughter cell will bud at the end or "pole" of the cell opposite the point at which it budded from its mother (called the distal pole), whereas the mother cell will bud at either its proximal pole or its distal pole (hence the term "bipolar" for this pattern).

In this chapter, we will discuss the reasons why budding patterns are regulated as they are and the mechanisms by which the axial and bipolar patterns are set up and maintained. We will then discuss the protein molecules that mediate the choice of budding site. Finally, we will learn how budding pattern is controlled by cell type.

Axial Budding Helps Haploid Cells to Mate after Mating-type Switching

Before delving into the "how" of budding patterns, I will first discuss the "why." For the haploid cell types, the axial budding pattern is thought to allow haploid cells to mate efficiently after mating-type switching (and, as mentioned in Chapter 6, **a**/α appears to be the preferred cell type state of yeast in the wild). Figure 8-1 depicts a cell that was a mother cell in a previous generation undergoing mating-type switching, a division, another round of mating-type switching (in the mother cell of course; see Chapter 6), followed by another division.

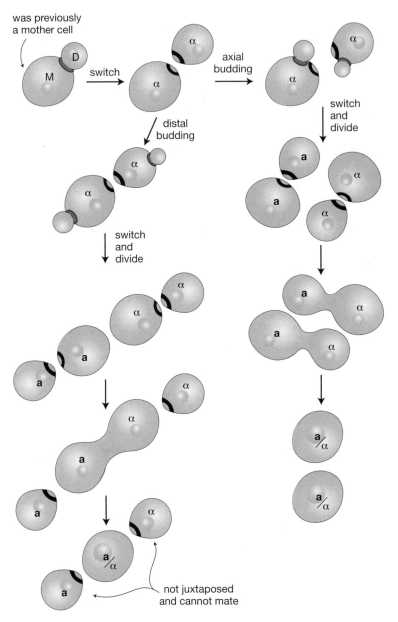

Figure 8-1. The pattern of budding influences the chances for mating success after mating-type switching.

Two scenarios are depicted in Figure 8-1. In the first, both the mother and daughter cell produced by the first division then bud at positions distal to their previous bud sites. The result of this would be four cells, arranged linearly, in which the first two cells (the ones on the left) would be mating type **a** and the last two (the two on the right) would be mating type α. In this case, the only two cells of opposite mating type would be the two in the middle. Consequently, only two of the four cells would be able to undergo mating. In the second scenario (which is what actually happens in haploid cells), axial budding between the two rounds of switching occurs (Fig. 8-1). In this case, the four cells produced are juxtaposed in a way that each has a neighbor of opposite mating type and each is able to mate.

Bipolar Budding Is Required for Nutrient Foraging during Diploid Pseudohyphal Growth

As mentioned in Chapter 2, diploid cells starved for a nitrogen source in the presence of an adequate supply of glucose elongate and form branching filaments. The bipolar budding pattern is necessary for the formation of branching filaments of cells during pseudohyphal development (Fig. 8-2). If axial budding were to occur here, a clump of cells, as shown on the left, would be produced in place of filaments. The right-hand side of the figure shows the consequence of three rounds of distal budding—the formation of branching filaments of cells. In reality, the pseudohyphal cells exhibit a bipolar budding pattern rather than the strictly distal pattern depicted.

Protein Landmarks Associated with the Plasma Membrane Are Required for the Control of Budding Patterns

Now that we know why yeast cells would like to control their pattern of cell division, we are in a position to consider how. In this section, I will describe the essentials of the processes that determine the axial pattern in haploid cells and the bipolar pattern in diploid cells. I will describe the proteins involved in the subsequent sections.

How the axial pattern is set up can be understood by first considering the following question: How might a haploid cell determine where it budded previously so that it could bud next to that site in the next generation? A simple way of accomplishing this task would be to leave behind some type of "landmark" at the site budding. Such a landmark could be made up of proteins that indicated to the mother cell where to bud in the first place. That landmark could then be "read" in the next generation. If the cell had a mechanism for recognizing the old landmark and placing a new landmark next to it, it would be able to direct budding to a particular site. This is precisely how the axial pattern is determined.

Figure 8-3 depicts this mechanism. The left portion of the figure shows a mother cell budding off a daughter. This mother cell chose the budding site shown because it recognized the landmark (shown as a gray ring associated with the plas-

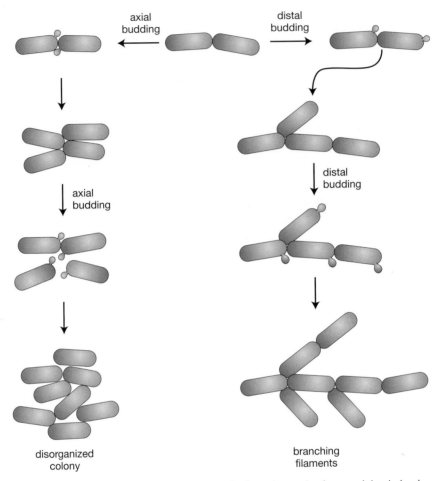

Figure 8-2. A distal budding pattern promotes the lateral growth of a pseudohyphal colony.

ma membrane) left behind from the previous cell cycle and assembled a new landmark (shown in black) next to it. This landmark determined the site of budding. In the next cell cycle, a new ring is formed (shown in blue) adjacent to the now-old landmark, which directs the mother cell to bud at that site in the next cell cycle (as shown on the right side of Fig. 8-3).

Although the daughter cell has never budded before, it uses a similar process to decide on its budding site. The daughter cell was born with a landmark (a "birthmark"), which was left at the point where it budded from its mother (shown in black in Fig. 8-3). Then the daughter cell assembles a landmark (shown in blue) adjacent to this birthmark, which determines where it forms a bud (as shown on the right side of Fig. 8-3).

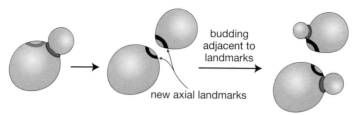

Figure 8-3. The axial budding pattern occurs because budding is directed to a site adjacent to an axial landmark deposited at the previous site of budding.

The bipolar pattern of **a**/α cells described at the beginning of the chapter is also directed by landmarks associated with the plasma membrane. However, there is one major departure from the process in haploid cells: **a**/α cells have three landmarks rather than one.

Figure 8-4 begins with a mother and daughter cell about to divide. For simplicity, we will say that the mother cell has never divided before (i.e., it was a

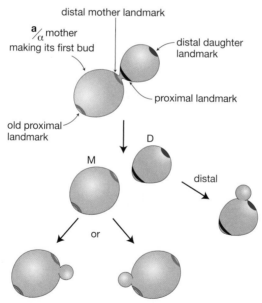

Figure 8-4. Bipolar budding mechanism. A newly born daughter (indicated by D) receives two landmarks, a "distal daughter" landmark and a "proximal" landmark. The distal daughter landmark acts in a dominant fashion to induce budding at the distal pole of the daughter cell in the next generation. The original mother cell (indicated by M) was a daughter cell in the previous generation. After budding, it contains two landmarks: the "distal mother" landmark that was deposited when it produced its daughter and the proximal landmark deposited in the previous generation (when it was born). In the next generation, the mother cell will bud next to one of these landmarks. Note that the proximal and the distal mother landmarks in the bipolar mechanism persist for many generations.

daughter cell in the previous generation). The daughter cell is going to be born with two different landmarks, the "proximal" and the "distal." The proximal landmark indicates where the cell budded from its mother cell (called "proximal" because it is laid down around the site of its birth just like the axial landmark in haploid cells). It also will contain a distal landmark at the opposite pole of the cell that was deposited there by its mother (Fig. 8-4).

Immediately after division, the mother cell will also have two landmarks. One was left from the previous division when the mother cell was itself a daughter cell (this landmark survives for many generations) and the other is a distal landmark that is deposited where it budded off its daughter (which also survives for many generations). As we will see below, the molecular composition of the distal landmarks of mother versus daughter cells are distinct. To indicate this difference, these are referred to in Figure 8-4 as the "distal mother" and the "distal daughter" landmarks, respectively.

How, then, do these landmarks dictate where budding will occur in the next cell cycle? The key is that if cells have the distal daughter landmark, then budding will occur at that site, no matter which other landmarks are in the cell. As a result, the daughter cell will always bud at the distal site in the next cell cycle (Fig. 8-4). The distal daughter landmark is destroyed in the process of budding. In a cell that was previously the mother cell, there are two landmarks as described above; one of these will be chosen at random to program the site of budding (Fig. 8-4).

BOX 8-1. PROPER CHOICE OF CELL DIVISION PLANE IS CRITICAL FOR THE GROWTH OF
EPITHELIAL TISSUES AND THE PREVENTION OF CANCER

Epithelial cells are polarized cells that line the lungs, intestines, and many other organs in the body. In general they have an apical surface, which faces the lumen of the organ in which they are located, and a basolateral surface, which forms an attachment to the surrounding tissue. The cells are fastened together by tight junctions that define the border between the apical and basolateral membranes. For an epithelial cell layer to expand, the plane cell division must be carefully controlled. As shown in Figure 8-5, division perpendicular to the axis of a sheet of epithelial cells maintains the sheet, but division along the axis of the sheet above or below the plane of the tight junctions results in one cell that stays in the sheet and another that is released from the sheet. The effect of the second plane of division is that (1) no net cells are added to the sheet and (2) a wayward epithelial cell has been created. The most common cancers are of epithelial cell origin and it is thought that their escape from the normal growth controls imposed by the epithelial environment is a critical step in their initiation.

(Continued)

Box 8-1. (*Continued*)

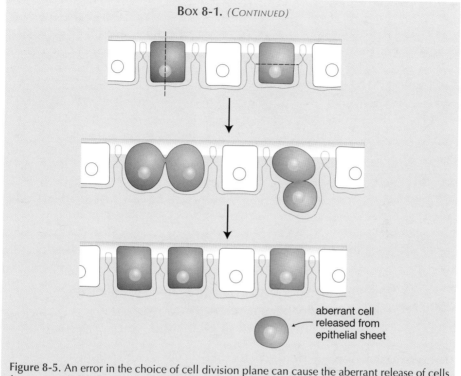

aberrant cell
released from
epithelial sheet

Figure 8-5. An error in the choice of cell division plane can cause the aberrant release of cells from an epithelial sheet.

The Axial Landmark is Composed of Four Membrane-associated Proteins

We will now consider the proteins that make up the landmarks discussed above. In haploid cells, the landmark is made up of many copies of a transmembrane protein named Axl2 and three associated cytoplasmic proteins, Bud3, Bud4, and Axl1 (Fig. 8-6). These proteins form rings that determine the site of division in both mother and daughter cells, and each progeny cell produced by that division then inherits these landmarks. Again, in the next generation, a new ring is formed adjacent to the landmark from the previous generation. How these proteins assemble into a ring structure and how the ring from the previous generation induces the formation of a new, adjacent ring are unknown.

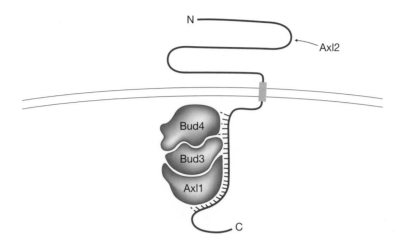

Figure 8-6. The proteins that compose the axial landmark.

The Proximal and Distal Daughter Bipolar Landmarks
Contain Transmembrane Proteins That
Are Related to Each Other

The proteins that make up the landmarks in **a**/α cells are distinct from those of the axial landmark (Fig. 8-7). Both the proximal landmark and the distal landmarks contain many copies of transmembrane proteins called Rax1 and Rax2. However, the proximal landmark also contains the protein Bud9, whereas the distal daughter landmark contains a protein called Bud8, and the distal mother landmark contains neither (it is hypothesized to contain as as-yet unidentified protein). Bud8 and

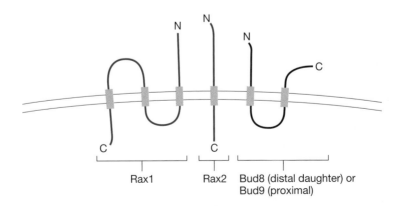

Figure 8-7. The proximal and distal landmarks of the bipolar budding mechanism share protein components.

Bud9 are homologous to each other, suggesting that they evolved from a single ancestral protein. The proximal landmark and distal mother landmark are assembled as rings around the site of budding (Fig. 8-4). In contrast, the distal daughter landmark that is laid down at the tip of a daughter bud is deposited there by secretory vesicles riding on actin cables (analogous to the delivery of Ash1 mRNA to the bud tip; see Chapter 6).

Nonrandom Budding Patterns Require a Small GTPase Cycle That Targets the Budding Machinery to a Specific Site

The landmark proteins do not interface directly with the machinery of the cell that generates a bud. Instead, there is a protein module that "plugs" into the various landmarks and then recruits the essential cell machinery involved in budding. By the term "module," I mean a group of proteins that function together and that can be recruited by other proteins in the cell to perform a particular task. The analogy in computer programming would be a function that can be called again and again by the main program when a particular task must be performed. The proteins that make up this module were discovered through genetic screens that identified mutants that caused both haploid and diploid cells to bud randomly (i.e., the site of budding in these mutants was unrelated to the previous site of budding).

The protein module is made of the proteins Bud1 (also known as Rsr1 for "ras-related"), Bud2, and Bud5. Bud1 is a small GTP-binding protein of the *ras* family that exists in two conformations, depending on whether it is bound to GTP or to GDP. Like other members of this family, it is anchored to the plasma membrane by a lipid modification of its carboxyl terminus. The interconversion of the two forms of Bud1 is catalyzed by the other two proteins of the module: Bud2, which induces GTP to be hydrolyzed on Bud1, and Bud5, which induces the exchange of GTP for GDP on Bud1 (Fig. 8-8). The landmarks discussed above cause Bud1 to cycle between these two states by themselves recruiting Bud2 and Bud5. The cycling of Bud1 between its two states causes budding to happen.

Bud1 binds to the proteins involved in budding, but, depending on its conformation, it binds to different proteins. In the GTP form, Bud1 binds Cdc42 and its exchange factor Cdc24 (we have already encountered these proteins and those discussed below in the context of shmooing, another form of polarized cell growth). In the GDP form, Bud1 binds to the protein Bem1, another protein required for both shmooing and budding. Repeated cycling between these two states allows one molecule of Bud1 to recruit many molecules of Cdc42, Cdc24, and Bem1 in an ordered fashion. As with shmooing, these proteins promote budding by recruiting additional proteins such as the formin Bni1. As described in Chapter 5, Bni1 directly induces the polymerization of actin into filaments, which in turn direct secretory vesicles to sites of cell growth.

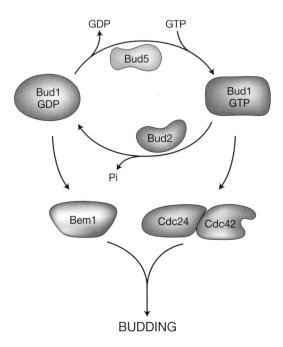

GDP   GTP

Bud5

Bud1
GDP

Bud1
GTP

Bud2

Pi

Bem1   Cdc24 Cdc42

BUDDING

Figure 8-8. A GTPase cycle induces the recruitment of many copies of Bem1, Cdc24, and Cdc42 at the site of budding. Recruitment of Bud2 and Bud5 by the protein landmarks described in Figs. 8-6 and 8-7 causes Bud1 to cycle between GTP and GDP forms. The GDP form of Bud1 recruits Bem1, whereas the GTP form recruits Cdc24 and Cdc42. These proteins in turn recruit proteins that build the incipient bud.

The Transcriptional Regulation of the *AXL1* Landmark Protein Explains the Regulation of Budding Patterns

The master regulators of cell type control the patterns of budding by a simple mechanism: *AXL1* is repressed in **a**/α diploids by the **a**1-α2 complex; without Axl1, the axial landmark does not form. In the absence of an axial landmark, the bipolar landmarks operate. Importantly, this also means that in haploid cells, both types of landmarks (axial and bipolar) are present, but the axial landmark is apparently more potent and therefore wins the competition for Bud2 and Bud5.

Review Article

Casamayor A. and Snyder M. 2002. Bud site selection and cell polarity in yeast. *Curr. Opin. Microbiol.* **5:** 179–186.

CHAPTER NINE

EVOLUTION OF CELL TYPE DETERMINATION

There is considerable evidence that all organisms evolved from a common ancestor that was the first life-form on Earth several billion years ago. It is thought that much of the great diversity that is found in modern-day organisms has arisen through changes in gene regulation. For example, differences in the morphology between organisms can be correlated with differences in the expression patterns of transcription factors that program the identities of body regions. The mating-type systems of fungi offer simple models for understanding changes in gene regulation through evolution. As discussed in Chapter 3, the identities of *Saccharomyces cerevisiae* cells are determined by the identities of the transcription factors expressed from the *MAT* locus. Two questions therefore arise: Is this circuitry set up the same way in different species and, if not, can the differences be understood in terms of their benefits to the organism? We are only beginning to understand evolution at the molecular level, and our answers to these questions are incomplete. In this chapter, we will discuss two other species of yeast to illustrate the natural diversity that exists in how cell type can be regulated. The first species we will discuss is distantly related to *S. cerevisiae* and has a complete sexual cycle. The other yeast is closely related to *S. cerevisiae* and appears to have lost the ability to undergo meiosis; although, under the right circumstances, it can mate.

In *Cryptococcus neoformans*, the *MAT* Loci Encode Pheromones and Their Receptors, Bypassing the Need for Haploid Mating Type–specific Transcriptional Regulators

Cryptococcus neoformans is a haploid yeast that causes disease in immunocompromised individuals. It is estimated to cause 15% of the deaths worldwide from HIV/AIDS. This fungus is particularly dangerous because it causes meningoencephalitis (inflammation of the brain and surrounding tissues), which, if left untreated, is always fatal.

    C. neoformans is found as a haploid strain of either mating type α or **a**. As in *S. cerevisiae*, these cell types can mate with each other to produce **a**/α cells, which then undergo meiosis and sporulation. *C. neoformans* is only very distantly related to *S. cerevisiae* and has a dramatically different mating-type locus.

    In *S. cerevisiae*, the genes encoding mating pheromones and their receptors are cell type–specific, as dictated by regulators encoded by the *MAT* loci. In *C. neoformans*, the expression of mating pheromones and receptors are also cell type–specific. For example, the α-pheromone receptor is expressed only in *MAT***a** cells. However, this pattern of expression is accomplished in a way completely different to the method used by *S. cerevisiae*: The pheromones and their receptors are encoded by the *MAT* locus itself (Fig. 9-1). Thus, *MAT***a** contains the genes for **a**-factor and the α-factor receptor, and *MAT*α contains the genes for α-factor and the **a**-factor receptor (Fig. 9-1). Because it uses this simple way of linking mating type to the expression of pheromones and their receptors, *C. neoformans* does not encode the *MAT* locus proteins that function in haploid cells to control expression of genes such as α1 and α2. Apparently, *C. neoformans* has given up the option of having **a**-sgs and α-sgs that are encoded outside the mating-type locus in favor of a more "hard-wired" connection between the mating-type locus and the genes encoding pheromones and receptors.

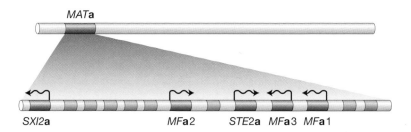

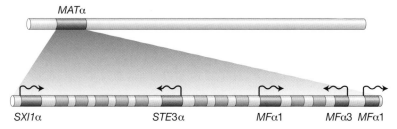

Figure 9-1. Mating-type locus of *Cryptococcus neoformans*. Note that multiple genes encode **a**-factor and α-factor.

Sxi2a-Sxi1α ⟶┤ h-sgs

Figure 9-2. Sxi2**a** and Sxi1α function in the **a**/α cell to repress haploid-specific genes.

The *C. neoformans MAT* Locus Encodes Homeodomain Proteins That Repress Haploid-specific Genes: Similarities to **a**1-α2

Despite its differences with *S. cerevisiae*, *C. neoformans* has maintained another aspect of mating-type control. The mating-type locus encodes one of two home-odomain proteins, which are analogous to **a**1 and α2. These are called Sxi1α and Sxi2**a** (Fig. 9-1). The corresponding genes are not transcribed in haploid cells. They are switched on only after **a** and α cells fuse during mating. In **a**/α cells, the expressed proteins are likely to form a complex, similar to **a**1-α2, that represses haploid-specific genes (Fig. 9-2). Presumably, one of these genes would be a repressor of meiosis, similar in function to the *S. cerevisiae* Rme1 protein, but such a factor has yet to be identified.

The Vast Majority of *C. neoformans* Isolates Are Strain *MAT*α—Does This Indicate the Loss of Sexual Reproduction?

Curiously, the two mating types of *C. neoformans* are not equally represented in the wild. Most isolates, from the environment and from patients, are mating type α strains; mating type **a** strains are quite rare. Because sexual reproduction produces an equal ratio of **a** and α progeny, the overrepresentation of mating type α strains in the wild suggests that *C. neoformans* rarely undergoes its sexual cycle in nature. Second, these observations suggest that the **a** mating type is less fit in some way. This could be due to the accumulation of deleterious mutations in genes important for optimal growth that have been trapped within the *MAT* locus (although this hypothesis remains to be proved). By "trapped" I mean that they cannot be sepa-rated through recombination from the mating-type locus because they are flanked by DNA sequences that are not homologous between *MAT***a** and *MAT*α. Thus, *C. neoformans* may represent a species in which sexual reproduction is on its way out or one that is in transition to a new mating-type system.

Candida albicans Is a Diploid Yeast with No Known Haploid Phase

Candida albicans will be familiar to many as the cause of oral thrush in infants and yeast infections in women. However, *C. albicans* can cause invasive infections of vir-tually any organ of the body in immunocompromised patients. *C. albicans* is a part of the normal human gut flora, but outside of humans and other animals, it is not found in the environment. This suggests it has become a specialized parasite of animals.

In the context of the entire fungal kingdom, *C. albicans* is quite a close relative of *S. cerevisiae*. It can grow as budding yeast that is indistinguishable from *S. cerevisiae* under the microscope. Like *S. cerevisiae*, it can switch to a pseudohyphal form of growth, but unlike *S. cerevisiae*, it can also switch to a true hyphal form of growth, producing long tubes with walls or septa between the cells. (Many fungi grow exclusively as hyphae, which results in the fuzzy appearance of colonies [e.g., bread mold].)

   C. albicans isolates from infected patients are always diploid organisms. Neither sporulation nor haploid forms have been observed, suggesting the organism may not have a complete sexual cycle that includes meiosis. However, *C. albicans* does have a mating-type locus. Although 99% of diploid isolates are **a**/α cells, 1% of isolates contain only the **a** or α version of the mating-type locus (yet are diploid—I will explain further below).

C. albicans Contains a Mating-type Locus Similar to That of *S. cerevisiae*, but Also Contains Genes Unrelated to Mating Type

The *MAT*-like locus in *C. albicans* is on chromosome V (Fig. 9-3) and its alleles are called *MTL***a** and *MTL*α (*MTL* stands for "MAT-*like*"). They are analogous to *MAT***a** and *MAT*α of *S. cerevisiae*. *MTL***a** encodes two proteins, **a**1 and **a**2. **a**1 is a homeodomain protein, homologous to *S. cerevisiae* **a**1, but the *C. albicans* **a**2 is a tran-

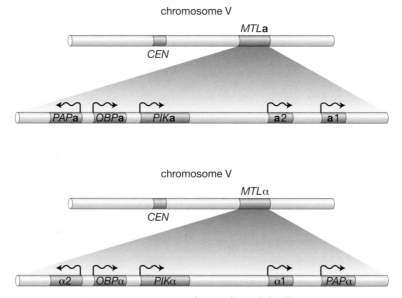

Figure 9-3. Mating-type locus of *Candida albicans*.

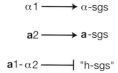

Figure 9-4. Function of regulators encoded by the mating-type locus of *Candida albicans*. Note that **a2** activates **a**-sgs and is unrelated to *S. cerevisiae* **a2**. Also note that α2 does not repress **a**-specific genes in *C. albicans*.

scription factor that is unrelated to the *S. cerevisiae* **a2**. *MTL*α encodes homologs of *S. cerevisiae* α1 and α2. In addition to the cell type regulators, the *MTL* locus also contains genes unrelated to cell type that appear to have been "trapped" during evolution (in the manner described above for the *C. neoformans MAT* locus).

Recall from Chapter 3 that in *S. cerevisiae*, **a**-sgs are constitutively active in **a** cells but are repressed by α2 in α cells. The *S. cerevisiae* α-sgs require activation by α1 to be transcribed and are therefore expressed in α cells, but not in *MAT***a** cells. The master regulators encoded by the *MTL* locus in *C. albicans* function differently from those in *S. cerevisiae* in two respects (Fig. 9-4): (1) *C. albicans* **a2** activates **a**-sgs, and (2) *C. albicans* α2 does not repress **a**-sgs (but still acts with **a1** in **a**/α cells to repress genes). The reason for the second difference is simple: The transcription of the **a**-sgs in *C. albicans* relies on the presence of the **a2** activator; because this protein is missing in α cells, there is no need for α2 to repress the **a**-sgs.

Studies of many related yeast species suggest that the common ancestor of *S. cerevisiae* and *C. albicans* had a *C. albicans*-like mating-type locus. During evolution of *S. cerevisiae*, it dispensed with the *C. albicans*-like **a2** gene and α2 evolved the ability to repress **a**-sgs (Fig. 9-4). Thus, *S. cerevisiae* accomplishes the same regulation using fewer regulators.

Loss of a Chromosome Containing One of the Mating-type Loci Is Required for *C. albicans* to Produce Mating-competent Cells

C. albicans has been isolated only in a diploid form. Based on what we have learned about *S. cerevisiae* in earlier chapters, one would expect these strains to be always of the **a**/α cell type. However, as mentioned above, about 1% of *C. albicans* strains are of either the **a** or α mating type, yet are diploid. Such **a** and α strains arise through the spontaneous loss (due to rare chromosome segregation errors during mitosis) of a copy of chromosome V that gives rise to cells that contain either *MTL***a** or *MTL*α, but not both (Fig. 9-5). Because cells lacking one copy of chromosome V grow poorly as a result of the twofold reduced expression of genes in chromosome V, cells that undergo a second segregation error event to produce progeny with two copies of chromosome V soon overtake the population. Such diploid **a** and α strains of *C. albicans* (which are diploid for all chromosomes) will mate with each other.

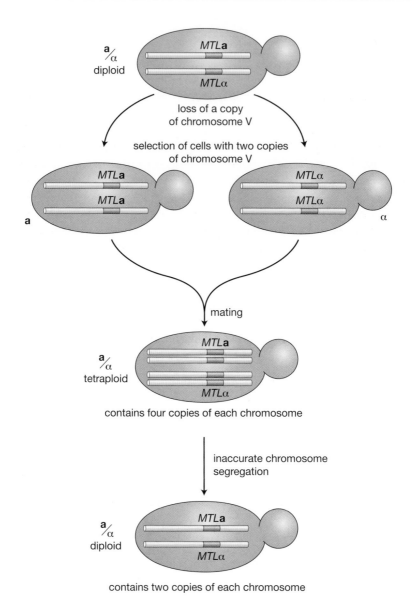

Figure 9-5. Current view of *Candida albicans* mating cycle.

Tetraploid Cells Are the Product of Mating: They Lose Chromosomes to Reestablish the Diploid State

The cells produced by the mating of *C. albicans* contain four copies of each chromosome (and are therefore called "tetraploid"). The tetraploid state appears to be unstable, and, through a process of chromosome loss during cell division, a population of tetraploid cells eventually becomes a population of diploid cells (Fig. 9-5). The *C. albicans* life cycle has been termed "parasexual" to indicate that it has only a partial sexual cycle. That is, although mating does occur, the subsequent reduction in chromosome number is achieved through mitotic rather than meiotic divisions. Of course, it cannot be ruled out that *C. albicans* can undergo meiosis under conditions that have yet to be identified. However, it seems possible that because it has become so well adapted to the restricted environment of its animal hosts, *C. albicans* no longer needs a complete sexual cycle to evolve in the context of its host.

In Contrast to *S. cerevisiae*, Mating in *C. albicans* Requires **a** or α Cells to Switch to a Different, More Vulnerable, Opaque Form

Early experiments suggested that mating between the **a** and α yeast cells of *C. albicans* was far less efficient than mating observed between the **a** and α cells of *S. cerevisiae*. Although this was initially thought to reflect poorly optimized laboratory conditions, further studies revealed a different reason for the apparent inefficiency. It transpires that the ordinary yeast form of *C. albicans* is actually unable to mate. Instead, in populations of *C. albicans* yeast **a** or α cells, there is a tiny fraction of cells that have switched to a different form (see Fig. 9-6) that can mate very efficiently. This mating-competent form is called the "Opaque" form (historically named as such because colonies of these cells are slightly darker than colonies formed by yeast cells; these cells differ from those in the White state in their shape, metabolism, and pattern of gene expression). Pure populations of **a** and α Opaque cells mate with each other with high efficiency. One of the reasons for this is that when cells switch from the White to the Opaque form, they turn on several genes required for mating.

The Opaque and White states are metastable. That is, Opaque cells tend to give rise to Opaque cells, but very occasionally give rise to White cells, which in

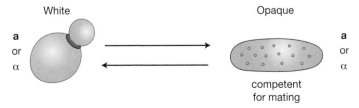

Figure 9-6. Switching to the Opaque state is required for *Candida albicans* to mate.

Figure 9-7. *Candida albicans* Wor1 is required for the switch to the Opaque state, activates its own promoter, and is repressed by the **a**1-α2 repressor.

turn tend to give rise to White cells. A positive-feedback mechanism has been identified that can explain the epigenetic inheritance of the Opaque state: A regulator called Wor1 (White–Opaque Regulator 1) that is essential for switching to the Opaque form binds to and activates its own promoter. Thus, cells that accumulate enough Wor1 to set up the positive-feedback loop will switch the Opaque state until the loop is broken by a reduction in Wor1 levels. The regulation of Wor1 also explains why **a**/α cells cannot switch to the Opaque form: **a**1 and α2 repress the transcription of *WOR1* (Fig. 9-7).

Why does *C. albicans* require that cells switch to a different form before mating, when *S. cerevisiae* and many other yeast species do not? The answer to this question is also uncertain. One idea is that because the Opaque form is more vulnerable to environmental attack, the White–Opaque switching mechanism might allow *C. albicans* to mate while maintaining a robust majority of White cells that will survive the hazardous conditions present in the host.

To summarize, the *C. albicans* mating-type locus functions differently from that of *S. cerevisiae* in two major ways. First, the transcription factors encoded by the *C. albicans* mating-type locus behave differently: **a**2 activates **a**-sgs, whereas α2 does not repress **a**-sgs. Second, **a**1-α2 represses a switching mechanism, the White–Opaque switch, that controls the ability of **a** and α cells to mate. These differences serve to illustrate how a transcriptional circuit can evolve over time from a common ancestor. Of course, to gain a better understanding of these evolutionary developments will require an in-depth analysis of the cell type circuitry in a large number of related yeast species. The hope is that we will ultimately be able to trace, in detail, the sequence of events that lead to the rewriting of transcriptional circuits as organisms evolve and to gain insights into the selective advantages afforded by these changes.

Review Articles

Bennett R.J. and Johnson A.D. 2005. Mating in *Candida albicans* and the search for a sexual cycle. *Annu. Rev. Microbiol.* **59:** 233–255.

Hull C. and Heitman J. 2002. Genetics of *Cryptococcus neoformans*. *Annu. Rev. Genet.* **36:** 557–615.

McClelland C.M., Chang Y.C., Varma A., and Kwon-Chung K.J. 2004. Uniqueness of the mating system in *Cryptococcus neoformans*. *Trends Microbiol.* **12:** 208–212.

INDEX